Technology *and* Religion

Templeton Science and Religion Series

In our fast-paced and high-tech era, when visual information seems so dominant, the need for short and compelling books has increased. This conciseness and convenience is the goal of the Templeton Science and Religion Series. We have commissioned scientists in a range of fields to distill their experience and knowledge into a brief tour of their specialties. They are writing for a general audience, readers with interests in the sciences or the humanities, which includes religion and theology. The relationship between science and religion has been likened to four types of doorways. The first two enter a realm of "conflict" or "separation" between these two views of life and the world. The next two doorways, however, open to a world of "interaction" or "harmony" between science and religion. We have asked our authors to enter these latter doorways to judge the possibilities. They begin with their sciences and, in aiming to address religion, return with a wide variety of critical viewpoints. We hope these short books open intellectual doors of every kind to readers of all backgrounds.

Technology *and* Religion

REMAINING HUMAN IN A CO-CREATED WORLD

Noreen Herzfeld

TEMPLETON PRESS

Templeton Press
300 Conshohocken State Road, Suite 550
West Conshohocken, PA 19428
www.templetonpress.org

Designed and typeset by Gopa & Ted2, Inc.

Library of Congress Cataloging-in-Publication Data
Herzfeld, Noreen L., 1956-
Technology and religion : remaining human in a co-created world / Noreen Herzfeld.
p. cm. — (Templeton science and religion series)
Includes bibliographical references and index.
ISBN-13: 978-1-59947-313-0 (pbk. : alk. paper)
ISBN-10: 1-59947-313-5 (pbk. : alk. paper)
1. Technology—Religious aspects. 2. Technology—Moral and ethical aspects. 3. Science and religion. I. Title.
BL265.T4H38 2009
201'.66—dc22 2008050718

Printed in the United States of America

09 10 11 12 13 14 10 9 8 7 6 5 4 3 2 1

Contents

Preface

TECHNOLOGY IS a rapidly changing field. Each day brings news stories of the development of new technologies. Yet the questions and concerns that surround these technologies are perennial. We ask today what is lost when our children turn to Facebook as a favored method of communication; a recent article in *The Atlantic* titled "Is Google Making Us Stupid?" questions whether the Internet is changing both the way we read and the way we think, making us less inclined to deep thought and reflection.[1] Plato asked the same questions 2,500 years ago, addressing the then-new technology of the written word:

> It will introduce forgetfulness into the soul of those who learn it: they will not practice using their memory because they will put their trust in writing, which is external and depends on signs that belong to others, instead of trying to remember from the inside, completely on their own. You have not discovered a potion for remembering, but for reminding; you provide your students with the appearance of wisdom, not with its reality. Your invention will enable them to hear many things without being properly taught, and they will imagine that they have come to know much while for the most part they will know nothing. And they will be difficult to get along with, since they will merely appear to be wise instead of really being so. (*Phaedrus* 275a–b)

The more things change, the more they stay the same.

Technology is also an extremely broad field. It is impossible in a single book to cover the gamut of new technologies and their implications for ourselves and our world. I have attempted in this volume to address a wide variety of technologies, dividing the field into three categories: technologies of the human body, technologies of the human mind, and technologies of the external environment.

I have written the bulk of this book while serving as a Fulbright scholar in Sarajevo, the capital of Bosnia and Herzegovina, a city that has historically inhabited the borderland between Christian West and Muslim East. As I type, I listen to bells from both Catholic and Orthodox cathedrals, ten minutes down the hill from my apartment, to be followed by the call of the muezzin from the five mosques I can see from my front porch. In the spirit of Sarajevo, I have considered the impact of our new technologies from Christian, Jewish, and Muslim perspectives. Many of the chapters of this book have been delivered as lectures at the Franciscan Seminary of Sarajevo and the Faculties of Philosophy and of Islamic Studies, both at the University of Sarajevo, and have benefited from comments and criticism from my students and fellow faculty members. I have also benefited from the insights of my colleagues at St. John's University; I would like in particular to thank Fr. Roger Kasprick, Nick Hayes, Bernie Evans, Ernie Diedrich, and Chuck Rodell for sharing their expertise on particular sections.

Who are we as creative creatures fashioned in the image of a creator God? How can we best channel our creativity in order to serve God and our neighbor and to exercise responsible dominion over nature? What are our responsibilities in a rapidly globalizing world? The technologies examined in the following pages will continue to change. But the questions they force us to ask will remain.

Technology *and* Religion

CHAPTER 1
Of Morals and Machines

FEW TECHNOLOGIES are simpler than a flashlight. But for rural Africa, the rechargeable flashlight—called a "solar"—has turned out to be a revolution. Much of Africa lacks sufficient electricity, and electricity is almost nonexistent in rural areas. In 2004, the United Nations estimated that only 20 percent of Africans (excluding Egypt and South Africa) have access to electricity, only 2 percent in most rural areas. This makes a flashlight a life saver. Peter Gatuoth, a Sudanese refugee, writes, "In case of thief, we open our solar and the thief ran away. If there is a sick person at night we will took him with the solar to health center." In African homes, children rely on the solar to study for school at night. It replaces the wood, charcoal, and kerosene that smokes up huts and causes respiratory problems. Solar flashlights do not address the entire energy puzzle for Africans, but they do show how a small technology can improve a slice of African life and do so without disrupting the environment or family traditions.[1]

China faces a similar power shortage for its rapidly industrializing and growing population. One solution to its energy crisis is a leviathan compared to distributing solar flashlights. On completion in 2009, the Three Gorges Dam on the Yangtze River will be the world's largest producer of hydroelectric power. The dam's thirty-four generators will pour out as much power as eighteen nuclear power plants. In all, China plans twelve such plants in the Yangtze basin. They are expected to reduce China's heavy dependence on coal, currently the source of 67 percent of China's electricity, and thus reduce a chief source of pollution. Unlike a flashlight, however, the price for this electricity will be high for the nation's traditions. As it approaches completion, Chinese officials admit that the Three Gorges Dam project may spawn more problems than it solves. Since its proposal, officials have known that the dam would displace a lot of people—1.3 million have already been relocated. It now appears that the dam will affect hundreds of thousands more. Ecological concerns abound. The dam seems to be destabilizing the surrounding hillsides, causing landslides and displacing farmers into higher regions, leading to further erosion. The reservoir will inundate villages and industrial sites, causing an accumulation of industrial toxins and human sewage in the water, pollutants previously flushed out by the moving river. The most lasting impact will be on the preservation of China's long and illustrious history. As many as 1,300 archeological sites will disappear forever once the reservoir fills. And the beautiful scenery of the Three Gorges will be forever changed. The Three Gorges Dam is a triumph of technology and yet a harbinger of a new set of human misfortunes.

From Africa to China, technology reveals its benefits and its costs. What we hear about most are the benefits. Those who hold a stake in the development or dissemination of a new technology, the engineers and entrepreneurs, and the politicians and journalists they influence, tend to focus almost exclusively on the benefits, often resulting in a breathless sort of boosterism, a promise that the technological future will be nothing but rosy. According to this view, technology can solve most of our economic and social problems. Technology brings order to the chaos of the natural world. It is the key component of progress. One of the chief oracles of this gospel of progress is Bill Gates, the brilliant founder and ex-chairman of Microsoft. The world is "getting better," he recently told world economic leaders. Women and minorities have advanced, life expectancy has skyrocketed, and more people have a democratic voice and freedom. According to Gates,

> These improvements have been triggered by advances in science, technology, and medicine. They have brought us to a high point in human welfare. We're really just at the beginning of this technology-driven revolution in what people can do for one another. In the coming decades, we'll have astonishing new abilities: better software, better diagnosis for illness, better cures, better education, better opportunities and more brilliant minds coming up with ideas that solve tough problems.[2]

Gates sees technology as the means to save the world. Like many who are uncritical of technology, he admits that current technologies have brought some social and economic challenges, yet he believes that solutions for these, and all other difficulties, can be found in further technological development.

Not all who are familiar with high-tech industry take as rosy a view. Bill Joy, the former CEO of Sun Microsystems, warned of the dangers of out-of-control technology in a 2000 *Wired* magazine article, "Why the Future Does Not Need Us." Joy is particularly concerned about the rapid development and convergence of robotics, genetic engineering, and nanotechnology. To the shock of *Wired* readers, he called for a moratorium. As he says,

> We are being propelled into this new century with no plan, no control, no brakes. Have we already gone too far down the path to alter course? I don't believe so, but we aren't trying yet, and the last chance to assert control—the fail-safe point—is rapidly approaching. . . . The experiences of the atomic scientists clearly show the need to take personal responsibility, the danger that things will move too fast, and the way in which a process can take on a life of its own. We can, as they did, create insurmountable problems in almost no time flat.[3]

Like many who focus on the downside of modern technology, Joy sees only problems ahead. He asks the impossible, that societies on a global scale agree to give up the short-term benefits of several technologies in order to safeguard the future.

Between these two extremes—that technology is entirely beneficial or entirely detrimental—is a third stance that is equally problematic. This voice declares that technology is morally neutral. The surgeon and Jack the Ripper give the knife its good or evil role—in short, a machine does not have moral agency. As the National Rifle Association slogan goes, "Guns don't kill people; people do." While this seems like common sense, the claim of moral neutrality is not entirely true. Modern technology does possess a certain amount of agency. Consider, for example, robotics and artificial intelligence. While a robot is indeed programmed by a human being, it is impossible for that human to understand and foresee each action the robot will subsequently take.

While this might be considered an extreme example, the problem casts a wide net. In a conflict situation, the presence of "neutral" guns does, in fact, increase the likelihood of death. The abundance of technology may also override many of our human choices. The New England transcendentalist Henry David Thoreau, writing in his cabin by Walden Pond in 1854, lamented: "We do not ride on the railroad, it rides upon us."[4] Thoreau chose not to ride on the railroad, but he could not remove the railroad—visible from his rustic hovel—from his community nor turn the landscape back to how it was before the tracks were laid. Similarly, we are only now seeing the social and environmental costs of the American love of the automobile.

Beyond environmental changes, every new technology also displaces an older one, often making the older technology no longer available. As Muslim philosopher Muzaffar Iqbal recently noted, "I cannot travel on camel to go to the haj as my grandfathers used to do."[5] This lament could sound sentimental; few really wish to return to an earlier time. But the point is valid enough. The losses caused by technology are usually seen only in retrospect, when we find that a new technology did not take into account all our values and intentions and the loss may be irreversible. For these reasons, technology is hardly a neutral matter. "It is a power endowed with its own peculiar force," writes Jacques Ellul, a French philosopher who has shaped much of the modern debate over technology. "It refracts in its own specific sense the wills which make use of it and the ends proposed for it. Indeed, independently of the objectives that man pretends to assign to any given technical means, that means always conceals in itself a finality which cannot be evaded."[6]

Technology is here to stay. We have always been creators of technology, from the first hurling of a stone at an animal in order to produce lunch or the first rubbing of two sticks or striking of flint against stone to create and harness fire to cook that lunch. We cannot escape technology. Yet we also should not blindly espouse every technology that comes down the pike. To choose which technologies serve us well and which do not, we need a way to assess them critically. The assessment must look at both the tool itself and the society in which it will be used. That, in turn, will require a clear grasp of who we are as individuals and societies and the values that we want to live by. This is where religion, a chief source of our cultures and values, plays an essential role in our discussion of technology.

Our religious communities preserve the wisdom of our forebears on questions of who we are and what we value. They also provide a locus for discussion in the context of worldwide communities of faith. Like technology, religion is also here to stay, despite the predictions of early twentieth-century intellectuals—Karl Marx, Sigmund Freud, Friedrich Nietzsche—to the contrary. In fact, as our devotion to science and technology grows, so it seems does our religious devotion. In the U.S., three times as many people regularly attend religious services today as compared to when the nation was founded. Technology and religion are growing hand in hand. Both form our worldview, our vision of who we are, our place in the world, and future trajectory. Thus, it makes sense to look at the two together.

What Is Technology?

Technology has one purpose—to change the world, to reshape ourselves or our environment. Through technology, we seek to stay safe from the elements, from predators, to make our lives longer and more comfortable. Yet modern technology goes beyond this defensive role. Through processes such as genetic engineering or nanotechnology, we seek not only to make our lives safer and easier but also to create things that are entirely new, in *Star Trek* terminology, "to boldly go where no man has gone before."

The word *technology* comes from the Greek word *techne*, which translates to mean craft, art, or knowledge. In other words, *techne* is about more than tools. Technology is not just the machines, chemicals, or instruments we use but also the techniques, processes, and methods by which we use them. But these processes do not exist in a vacuum. Technologies define a large part of the society that uses them while they are at the same time developed by that society. Thus, *techne* has three elements to it—tools, processes, and a social context.

On first glance, *techne* and modern technology look much the same, yet modern technology is substantially different from the *techne* of the ancients. Social critic Neil Postman emphasizes this distinction in his popular work *Technopoly: The Surrender of Culture to Technology*. Unlike the ancients, he says, modern Western culture has engaged in "the deification of technology, which means that the culture seeks its authorization in technology, finds its satisfactions in technology, and takes its orders from technology."[7] The tools of *techne* solve pressing physical problems in the production of food, clothing, and shelter. They aided in our ancestors' survival but did not play a major role in their understanding of the world and their place in it. These tools were not central to the thought world of the Greeks. In today's world, technology is central to our understanding of ourselves and the environment around us. Postman believes technology has, in fact, come to monopolize that understanding. Given the tremendously large role religion and religious identity still play in the modern world, I would suggest that our technological worldview has no monopoly but stands together, sometimes in tension, sometimes in harmony, with our religious understandings. But technology plays an undeniably greater role in our lives than it has at any previous time in human history.

That greater role is also seen in the power to create something new, a quest that was less prominent in ancient *techne*. To create the new is to go outside of nature. In his essay "The Question concerning Technology," the German existentialist Martin Heidegger observes that the ancient craftsman certainly made something new when he constructed a chair. A doctor might bring new health to a patient. However, neither imposed a new form on nature; rather, each worked with what is already implicit in the wood or the body. The wood of the chair is still wood and will rot just as would a log in the forest. The body returns to the natural healthy condition. In contrast, a genetically engineered human or a chimera that is half sheep and half goat is outside of its natural order. By its genetic alteration, the "geep" will never again produce a sheep or goat. Humanly extracted plutonium never returns to the uranium from which it was derived. The new products of modern technology do not simply "disclose" or shape nature but transform and replace nature.[8] In this way, modern technology gives us heretofore undreamed of power.

Faced with these prospects, it behooves us to bring the best of our wisdom to bear in selecting and using technologies. The French thinker Ellul, a critic of what he calls the modern technological tyranny, has composed a list of "76 Reasonable Questions to Ask about Any Technology" (see appendix). The list includes ecological, social, practical, moral, ethical, vocational, metaphysical, political, and aesthetic considerations. They are insightful questions, and well worth consideration. But few are likely to take the time to consider such an extensive list. Rather than add a tenth category of religious questions, we can find in our religious traditions a shorter list of more general concerns, a list that subsumes many of Ellul's categories.

The Idea of Creation

A commonality in how the monotheistic faiths—Judaism, Christianity, and Islam—approach technology springs from the book of Genesis, where the human project is so poetically introduced. In the first chapters of this book, we find two stories of creation. These stories attempt to teach a basic understanding of who we are as humans and what our place is in relationship to the rest of the created world. They are also stories about the act of creation, of bringing order out of chaos, and technology is at root a creative activity, one that brings order to the material world. Thus, there is no better place to begin a consideration of technology in a religious light than "in the beginning."

The creation of human beings is described in Genesis 1:26–28:

> Then God said, "Let us make humankind in our image, according to our likeness; and let them have dominion over the fish of the sea, and over the birds of the air, and over the cattle, and over all the wild animals of the earth, and over every creeping thing that creeps upon the earth." So God created humankind in his image, in the image of God he created them, male and female he created them. God blessed them, and God said to them, "Be fruitful and multiply, and fill the earth and subdue it; and have dominion over the fish of the sea and over the birds of the air and over every living thing that moves upon the earth."

These verses present both God's intent and action in creating human beings in the divine image. The expressions "in our image" and "according to our likeness" are used for no other creature. Humans are not merely the most developed form in God's creation but are set apart, relating to the divine sphere as well as to the created world. How we relate to the Divine is not explained outright, but verses 26–28 offer a clue. First, the inclusion of "male and female" in verse 27 suggests that both men and women participate in the divine image and that the divine image has something to do with us collectively, not merely as individuals.[9] Furthermore, the concept of dominion over the rest of creation appears twice—in God's resolution to create humans in the divine image in verse 26 and again, as a command, in verse 28.

These verses in Genesis begin to define our relationship to technology. The God presented in Genesis 1 is a creator. As creatures bearing the image of that God, humans are also creators. This precludes taking a totally negative view of technology. We are creators just as our God is a creator. Second, our creation of tools and methods with which to alter our environment should not be considered negative, for we understand ourselves to have been given dominion over the rest of creation. One of the great Old Testament scholars of our time, Gerhard von Rad, has delved into the nuances of the Hebrew words in Genesis and come back with a rich set of implications for the ideas of "image" and "dominion." He writes that being in the likeness suggests humans are "invested with might" in the world. "There is no other evidence in the [Old Testament] as to the proper interpretation of the divine likeness," he says. Humanity is given the divine image so that "[they] may control the whole of creation."[10] The Hebrew word for image, *selem*, connotes a material image and elsewhere in the Old Testament can be translated "duplicate," "idol," or "painting."[11] As von Rad summarizes,

> Just as powerful earthly kings, to indicate their claim to dominion, erect an image of themselves in the provinces

> of their empire where they do not personally appear, so man is placed upon earth in God's image, as God's sovereign emblem. He is really only God's representative, summoned to maintain and enforce God's claim to dominion over the earth.[12]

This same usage can also be found in other literature from the region. The Egyptians, in the seventeenth century BCE, used "image of God" to mean that the king is a living statue of god.[13] Mesopotamian literature from this period also contains several references to various Assyrian kings as being in the image of one or another of the gods in their exercise of power over the people.[14] The writer of Genesis 1 differs from this tradition by extending the image to all persons, so that the exercise of sovereignty becomes a universal trait for all human beings, not just the king. If all humans exercise sovereignty, then that sovereignty cannot be merely over other humans but over the rest of creation.

As represented by the work of von Rad, biblical scholars today generally agree that these Genesis texts describe men and women as God's deputies on earth. This approach has three persuasive strengths. First, it emphasizes a holistic view of human beings. The image of God is not found in merely the intellectual or spiritual capacities of the human; it is the whole of the human being, both physical and intellectual that exerts dominion over the earth. Von Rad warns against too sharp a split between the physical and spiritual, for "the whole man is created in God's image." As a second point, he shows that humanity "in the image" of God is declared in the context of a commission to exercise dominion.[15] The two go together, in other words. The third strength of this view hinges on von Rad's insistence that terms such as "dominion" be taken in the historical and literary context of antiquity. Modern ecologists have treated "dominion" with harsh skepticism, but von Rad points out that the writers of Genesis had an entirely different worldview. In composing Genesis 1, they viewed chaos as "the great menace to creation" and thus a threat to humanity. Humans are called upon to join God in imposing order on nature, a nature created in reference to humans. In doing so, humanity participates in God's saving plan.[16] Developing technology is part of this participation.

This kind of literary exegesis of Genesis in its original Hebrew, and in its historical context, has been extended by other theologians. While von Rad focused on understanding dominion, the systematic theologian Karl Barth has looked at two other parts of Genesis 1 in order to understand something about the relationships among God, humanity, and nature. Barth focuses on verse 26 ("Let us make man in our image") and verse 27 ("male and female he created them"). Barth notes the use of the plural in these verses. For example, he interprets "Let us make man" in a Trinitarian way, whereas others have seen the "us" as members of God's heavenly court (angels, for example). Barth reasons that God must be addressing his own self when it comes to so great an event as the primordial act of creation. The plural pronoun, he says, points to a triune Godhead that contains more than one person, a God who embodies relationship in God's very being.[17] This relational nature, existing within the Godhead, forms the ground of human creation. For Barth, and for many other theologians, human nature is typified by being in relationship, as first realized in the triune God.

This relationship can take two forms. Barth finds the first image of the relational Trinity in the human–God relationship:

> In God's own being and sphere there is a counterpart: a genuine but harmonious self-encounter and self-discovery; a free co-existence and co-operation; an open confrontation and reciprocity. Man is the repetition of this divine form of life; its copy and reflection. He is this first in the fact that he is the counterpart of God, the encounter and discovery in God Himself being copied and imitated in God's relation to man.[18]

The copy is in the relation itself, not in our capacity for relationship. Thus, the image of God in humanity is not a quality for Barth. It exists in particular relationships, first with God and then with one another. According to Barth, "Image has double meaning: God lives in togetherness with Himself, then God lives in togetherness with man, then men live in togetherness with one another."[19]

Barth also turns to the text "male and female he created them." He finds in this relationship the same dynamic as within the triune God. As God conceived it, Barth says, the male–female relationship is "the great paradigm of everything that is to take place between [man] and God, and also of everything that is to take place between him and his fellows."[20] In the Godhead, there is an "I" and a "Thou" in confrontation and conversation. Similarly, the true human being does not exist in isolation, but in encounter with another.[21] Barth sees the "male and female" of Genesis as a mandate for our necessary confrontation with people who are like us and different from us.[22]

In addition to the dynamic of male and female, this confrontation between difference and unity is especially strong in the God–human relationship. For Barth, God is "wholly other" to humanity. God created in order to enter into relationship with a being "which in all of its non-deity and therefore its differentiation can be a real partner; which is capable of action and responsibility in relation to Him."[23] This disjunction, between the infinite and finite, is not the case between male and female. Hence, Barth highlights the male–female difference in Genesis as a model for the differences we encounter in other humans. He describes it as the entire "dialectic" of human interaction, a daily experience "of gift and task, of need and satisfaction, of lack and fulfillment, of antithesis and union, of superiority and subjection."[24]

Barth moves on to Jesus as another exemplar. Jesus actively gives himself to his fellow humans: "If we see Him alone, we do not see Him at all. If we see him, we see with and around Him in ever widening circles His disciples, the people, His enemies, and the countless multitudes who never have heard His name. We see Him as theirs, determined by them and for them, belonging to each and every one of them."[25] He describes Jesus as "man for God and God for man."[26]

What we learn from von Rad and Barth is that relationship and dominion must work together in a creative dialectic. A brief look at Genesis 4–9 shows what can happen when human beings exercise dominion without considering the consequences for either human–human or human–God relationships. The age-old tension between the pastoralist and the user of the newer technologies of cultivation is explored in the story of Cain and Abel, where the relationship between brothers is shattered. The technologies of construction and of the city produce the tower of Babel, and whole societies cease to understand one another. Noah and his family are saved by the technology of the ark. Noah succeeds with technology where Babel failed. Noah couples the technology of the ark with relationship. He enters into a covenant with God, a covenant that repeats both the claim that humans are in God's image and that they have dominion over the earth. These chapters provide a warning, that humans will come to disaster should they attempt to "go it alone." Dominion without a relationship to God, to one another, and to the rest of creation can produce unforeseen and disastrous results.

We find a similar picture in Islamic thought. Islam does not specifically state that humans are created in God's image. Indeed, this thought would be considered blasphemous since God is wholly other and nothing can be compared to him. However, the creation of human beings is a special event in Islam, separate from that of the animals, and we find an echo of the theme of dominion in the Qur'an's treatment of the creation of humankind. The Muslim understanding of humanity's place as God's *khalifa*, or vice-regent, closely echoes that of von Rad. According to the Qur'an, the Lord addresses the angels, saying, "I will create a vice-regent on earth." The angels, a bit worried, reply, "Will You place therein one who will make mischief therein and shed blood? — while we do celebrate Your praises and glorify Your holy (name)?" The Lord then says, "I know what you do not know" (2:30).[27]

The Qur'anic vice-regency, like that of Genesis, is over nature, not over peoples, and is a special prerogative of humanity. "We did indeed offer the Trust to the Heavens and the Earth and the Mountains, but they refused to undertake it, being afraid thereof; but man undertook it" (33:72). Thus, Islam views technology as an exercise of human dominion. The Qur'an notes that nature is in part for humanity's use: "It is Allah Who has subjected the sea to you, that the ships may sail through it by His command, that ye may seek His bounty, and that you may be grateful. And He has subjected to you, as from Him, all that is in the heavens and on earth: behold" (45:12–13).

The Qur'an also notes that human beings can misuse their authority as *khalifa*. Surah 33:72, in which humankind accepts the trust declined by nature and the angels, continues with the observation "but man undertook it; — he was indeed unjust and foolish." The contemporary Islamic scholar Seyyed Hossein Nasr explains the continuing concern about human dominion, which we heard from the Qur'anic angels earlier: "Nothing is more dangerous for the natural environment than the practice of the power of vice-regency by a humanity which no longer accepts to be God's servant, obedient to His commands and laws."[28] Indeed the disastrous stories of Cain and Abel and the Flood appear in the Qur'an, much as they do in Genesis.

Humans are expected to execute their vice-regency with truth, justice, and compassion. This moves us into the realm of relationship. As in Genesis 1, the Qur'an also mentions the creation of both men and women: "O mankind! Reverence Your Guardian-Lord, Who created you from a single person, created, of like nature, his mate, and from them twain scattered (like seeds) countless men and women" (4:1). It reminds us (in surah 2:213) that we are essentially one people. Thus, the Qur'an is deeply concerned with relationship. It provides direction in a variety of places for how to live out our relationship with God, with one another, and with the natural environment. If there is a difference between the Christian and the Muslim understanding of this relationship, it is found in the different levels of specificity. The Qur'an provides specific roles and duties, specific responsibilities that guide human relationships, while Christians read simply that we should "love one another," the Gospels providing stories that explicate such love in action.

Making High-tech Choices

In the rolling farmlands of southeastern Pennsylvania, the residents have an electrical grid that would be the envy of both Africa and China. However, Pennsylvania's Amish community has decided to deal with modern technology by their own lights—their own system of evaluation. This historical group of Pietist immigrants, beloved in American culture for their simple lives, broad hats, and horse-drawn buggies, is hardly typical of how the vast majority of Americans want to live. But the Amish are a striking example of a group that interprets technology through a religious tradition.

Unlike most Americans, who feel expected to adapt to every new technology, the Amish have devised a way of choosing which technologies to adopt and which ones to reject. For them, technology comes second to both religious and cultural identity. The benefits of any technology are weighed against the changes it might bring to these two aspects of their community.

Contrary to the popular conception that the Amish have stopped the clock, using only technologies that were available in the nineteenth century, the Amish do adopt some modern technologies. For example, the Amish will use a telephone; they simply prohibit having one in each house, since this would interfere with the custom of visiting one another and discussing things face to face. The Amish use refrigerated milk tanks, powered by gasoline generators, since these are necessary if their milk is to be sold to the wider community, but they see no need for a refrigerator in every kitchen. Some even use solar panels to generate enough electricity to power a saw or a sewing machine, but they stay off the electric grid. In general, the Amish ask three questions when considering a new technology.

First, does the technology provide tangible benefits to the community or individuals within that community? The principal benefits the Amish look for are economic or medicinal. Modern antibiotics and diagnostic or surgical techniques are acceptable. So is the use of mowers, hay balers, or generator-powered woodworking equipment. Each contributes to the health or economic strength of the community. On the other hand, equipment used primarily for entertainment or ease is rejected as frivolous and unnecessary.

Second, does the technology change the relationship of the individual to the community? Changes that might cause pride or attract attention are unacceptable. Thus, plastic surgery would be forbidden, as is individual ownership of items like cars, computers, or anything the rest of us would consider status symbols. For the Amish, social equality is an important part of maintaining a harmonious community.

Third, does the technology change the nature of the community itself? In particular, the Amish wish to safeguard the constant face-to-face contact among community members that has fostered solidarity and cultural identity. Changes that significantly enlarge the radius of one's life are rejected as inimical to the life of the local community. Private cars, telephones, and e-mail would change the nature of community contact and weaken local ties. The Amish also look askance at technologies that would open the community to too much outside influence. Television and the Internet are seen as conveyors of the majority culture that pose a threat to both the values and the cultural identity of the Amish.[29] Similarly, anything that would seem contrary to the religious order of the community is immediately rejected.

While most modern Americans view the Amish approach to technology as extreme, the questions the Amish ask of a new technology are not unreasonable ones. The acute awareness the Amish have of the impacts of technology on society is a lesson for us all. The three questions asked by the Amish also return us to the points the biblical scholars extracted from Genesis 1. First, does the technology exercise human dominion in a useful and responsible way? What are the tangible benefits? These are usually quite obvious since most technologies are developed because of their tangible benefits. Does it heal the sick, help the poor and suffering, or make the environment a more sustainable and beautiful place?

The pitfalls described in Genesis 4–9—where technology plays a role in fratricide and national conflicts—bring up the question of technological effects on our relationships. Does it change the way we think of ourselves or what it means to be human? How will that change affect our relationship with God or with one another? Are there issues of equity and social justice—is it available to all or only to some? While these questions are easy to ask, the answers are not always obvious. The following chapters will raise these questions in light of very specific technologies of our age, ones that enter our lives intimately and that steal newspaper headlines almost daily.

We may not foresee the long-term effects on nature or on the human community when first considering a new technology. Even if we do see what look like inevitable results, we ourselves may or may not be able to exercise personal agency regarding a particular technology. I can choose whether to undergo certain genetic tests or to take a certain drug, but I cannot, as an individual, stop the release of nanoparticles into the environment or necessarily choose whether the foods I eat or clothes I wear come from genetically modified crops. I can choose not to play video games, but it would be difficult for me to choose not to use a computer and simultaneously maintain my current profession. Most choices about technology are corporate ones, made by a community rather than by individuals. And this community is increasingly global in scope.

As I hope to show in these pages, we are not powerless in the face of technology's advance. We can use the concept of relationship to temper the dominion gained through technology. With the dialectic of dominion and relationship in hand, I now turn to an examination of particular technologies. Our journey will take us through three territories where technology affects our lives and world. The first, in chapter 2, considers "technologies of the body," which range from genetic engineering to pharmaceuticals. Next come "technologies of the mind," by which I mean the effect of computer technology on our perceptions of intelligence and reality itself. In chapter 4, we will look at "technologies of matter," the ways we seek to alter our environment for our own purposes.

With each technology we meet, I will consider how it extends our dominion in the world and what benefits it offers. Then, I will look at how the technology changes the way we view ourselves as individuals, the way we view ourselves as a community, and the way we relate to one another. The book concludes with a look at how religion is responding to the globalization of technology and how this modern challenge of global relationships bears a new kind of awareness and responsibility. This is a fact-filled book, moved along by stories, case studies, headlines, and some of my own personal anecdotes. Knowing the facts is important. When we speak with facts at hand, especially with the collective voice of our particular ethical or religious communities, we gain the power to guide the invention and uses of technology in the future.

CHAPTER 2

Healing or Enhancing?

IF YOU HAVE picked up a newspaper in the last year or two, you will have read about a variety of amazing medical advances. New medications delay the onset of AIDS for up to twenty years. New prostheses replace lost limbs or even make it possible for quadriplegics to move a computer cursor with their thoughts. The recent completion of the Human Genome Project is allowing researchers to locate genetic markers for a variety of illnesses, including muscular dystrophy, Down syndrome, sickle cell anemia, Huntington's disease, and certain forms of cancer. Stem cell research may get a boost from a recently publicized breakthrough using skin cells rather than cells from frozen embryos.

Yet, over the same year or two, in the same papers, one could also find a darker side to these new technologies of the body. The Tour de France remains ridden with drug scandals. Runner Marion Jones was stripped of her Olympic medals, while the all-American sport of baseball has been plagued with one revelation after another of illegal steroid use. Meanwhile, pictures of genetically engineered animals—half goat, half sheep, a rabbit that glows in the dark—raise the specter of a future ability to alter irrevocably any species, including human beings.

Modern medical technology is Janus faced. It has considerably lengthened our lifespan and holds the promise of therapies or cures for a variety of illnesses we have heretofore found intractable. Yet it also provides the means for becoming "better than well," for continually raising the bar on human performance by enhancing our abilities; beyond that, it raises the prospect of changing our very nature. How are we to judge the good uses of this technology from the bad?

Everything about our cultures and our sacred books urges us to heal. For Christians, the New Testament Gospels speak of how frequently Jesus engages in healing the sick. The disciples are also sent to be healers. One of the sayings of Muhammad similarly notes that each illness has its treatment and that we should look for it: "Seek treatment, because Allah did not send down a sickness but has sent down a medication for it."[1] Clearly, none of us wishes to suffer or to see a loved one suffer. In the past, the treatments available to us were limited. They still are, yet we have pushed the boundaries of those limits in such a way that new questions arise. Is all suffering bad? Would the prospect for human creativity flatten out if we no longer experienced illness or pain? What does perfection mean for the human body? With these questions in mind, we can look at some of the most cutting-edge medical technologies.

Genetics, Stem Cells, and Cloning

On a rainy day in the late summer of 1980, Heather Summerhayes sat by her ailing sister, Pam, as a gray light came into the room through the drawn curtains. As Heather remembers the day, she felt as if they were "cocooned by death and the sound of heavy rain":

> My sister tries to breathe. Her lungs make crackling sounds as she takes quick, shallow gasps, her chest heaving in short bursts followed by long drawn out moments of utter stillness, each one longer than the last. My heart jumps before she breathes again. Her lips are blue. They flutter almost imperceptibly beneath the plastic of her oxygen mask. Her eyes are closed. My own breath comes hard against the pain of losing her. I hold her hand, sit-

> ting next to her bed in the room where she once dreamed her teenage dreams.[2]

Pam died of cystic fibrosis that year at the age of twenty-six. Who would not welcome a cure for an illness such as this? Cystic fibrosis, a disease in which the body produces excessively thick mucus that clogs the lungs and obstructs the pancreas, is one of many conditions caused by a defective gene. Other illnesses caused by a single defective gene include muscular dystrophy, Down syndrome, sickle cell anemia, and Huntington's disease. Autism, Alzheimer's, multiple sclerosis, and Parkinson's also seem to be genetically linked.[3] Even the most common diseases—diabetes, heart disease, cancer—are frequently accompanied by genetic alterations.

The technologies of genetic engineering, stem cell research, and cloning offer the dual possibility of prevention or cure of much that currently shortens our life or degrades its quality. Yet if we have the technology to alter genes that make us ill, why not alter genes that make us ugly, short, fat, or slow? In other words, the technologies we use for healing can also be used for enhancement. They could make our bodies stronger and our brains more powerful. Why not engineer your future child to be 6'5" tall and a star on the basketball court? Or choose the facial features and bone structure of a supermodel for your daughter? Why settle for playing "Baby Einstein" tapes for your child when you could make sure he or she is born with a genius IQ? The same technologies that hold out the promise of alleviating the suffering of so many of us also hold the possibility of changing the norms of what it means to be human.

Genetic Engineering

Genetic engineering per se is not a new technology. Farmers have long bred specific plants and animals to enhance particular traits. For example, selective breeding has given us more than two hundred breeds of dog, many with traits specifically chosen for particular activities such as hunting weasels or herding sheep. Gregor Mendel explained the basic laws of heredity in his 1866 paper, "Experiments in Plant Hybridization." There are, however, two problems with selective breeding as a method of enhancing or eliminating specific genetic qualities.

First, it is a hit-or-miss operation. One may need many, many iterations before a particular quality appears or disappears, and in targeting one quality one may also introduce unintended side effects. Breeding also works on aggregate communities. One may have to grow thousands of roses before one gets exactly the shade desired. You're not going to get what you want in the first try. But we want our child to be immune from Alzheimer's or cured of muscular dystrophy, not someone's great-great-great-great- . . . great-grandchild.

A second problem is that, while one may control several generations of dogs or roses, it is another matter to try to control human reproductive choices. We make those choices for a variety of reasons far from the genetic. While there have been some changes in the human gene pool over the centuries, those changes were never specifically planned.

To make instant and sure changes, we need to know exactly what gene is responsible for the trait desired and have the ability to copy, add, or delete genes as needed. In 1990, the scientific quest to locate specific genes began in earnest with the Human Genome Project. Using the newest technology, the project's goal was to produce a "sequence" or map of every gene on the forty-six chromosomes that make up the human genome. The sequencing was done by round-the-clock running of computers and automated machines that cut, dyed, and read the patterns in the genes, then put the pieces back in order again. The completion of a "first draft" of the human genome sequence was announced in 2000 by President Bill Clinton and Prime Minister Tony Blair, reflecting the international cooperation in the project.[4]

The project identified the approximately 3 billion base pairs (made of deoxyribonucleic acid or DNA) that comprise the human genome. These pairs make up the 20,000–30,000 genes that produce the proteins that allow our cells to function.[5] We now have a map of the molecular structure of our genome. This does not mean that we have a complete list of genes, nor do we know what each gene does or does not do. However, many variant genes that contribute to human illness have been identified. As a result, scientists have developed screening procedures for a variety of disorders. These include cystic fibrosis, Down syndrome, Huntington's disease, sickle cell anemia, Tay-Sachs disorder, and a predisposition to heart disease or breast cancer.

For actual treatment of these diseases, however, we need the ability to delete defective genes and add a copy of the proper gene to a person's cell. Germline genetic therapy makes this substitution in the egg or sperm cell, which would correct a genetic abnormality in all subsequently produced cells of the person's body, including his or her sperm or egg cells, ensuring that the trait corrected for would not be passed on to subsequent generations. To date, germline therapy is beyond our capabilities. All gene therapy has been somatic therapy, conducted on the adult cells of a living person and, hence, confined to those cells treated. The idea is to place the proper gene into a carrier or vector, usually a virus, that then invades the cell and deposits the gene. The new gene then sets up shop producing the proteins that may be missing. In principle, somatic therapy works best for diseases caused by a missing protein from a single gene. To be successful, the target cells must also be long lived; otherwise, the procedure would have to be repeated multiple times. Rapidly dividing cells have not proven to be stable enough to incorporate and propagate the new gene.

While experiments in animals have been promising, somatic gene therapy has not been very successful so far in human beings. In 1999, eighteen-year-old Jesse Gelsinger died from multiple organ failure while participating in a gene therapy trial. This failure seems to have been caused by an immune reaction to the carrier virus and illustrates one of the main problems with somatic gene therapy.

Our immune systems are programmed to respond to any recognized foreign material; thus, an immunological response is likely to the viral vector. Of course, the virus itself can also cause disease or toxicity. This happened to two children in France who developed leukemia while undergoing a trial therapy for an immune deficiency. Given these difficulties, the U.S. Food and Drug Administration (FDA) has currently halted all experimental trials in the United States using retroviral vectors in blood stem cells and has yet to approve any gene therapies for the market.[6]

For a lasting solution to a genetic defect, germline therapy holds more promise. By aiming the gene therapy at the sperm or egg cell, the correction is permanently anchored in the hereditary line. With a hereditary disease such as Huntington's, germline is the only effective remedy. Germline therapy is preventative medicine, keeping the disease out of future generations, and so would be much more cost effective in the long term than disease management.

More than most new medical technologies, however, germline therapy raises a host of troubling ethical questions. This would be our most direct power to alter heredity. The outcome in future generations could have effects that were unforeseen or unintended. And we cannot obtain the consent of our potential children and grandchildren to be experimented upon. Furthermore, all genetic therapies are currently extraordinarily expensive. At least in the foreseeable future, the use of germline technology is likely to follow the divisions of economic classes, both for medical purposes and for enhancement of our children. And if everyone could afford it, this technology could lead to a limiting of the diversity of the human gene pool, a future Lake Wobegon where everyone is strong, good looking, and above average.[7]

What about genetic Frankensteins? Scientists have already used germline techniques to create "chimeras," or mixed creatures, in their laboratories.[8] In 1984, a "geep" was produced by combining genetic material from a goat and a sheep. The genes that cause phosphorescence in certain varieties of fish have been inserted into rabbits and mice; they now glow in the dark. Rabbits and mice have also been given human genes so they can be subjects for testing the effect of drugs on the human immune response. So far, the science has only involved transferring bits of human genes into laboratory animals—and has raised few ethical or religious questions. The possibility of transferring animal genetic material into a human being, however, is a different story. It raises the question of what it means to be human. At what point would a transgenic human being cease to be "human"? When she glows in the dark? When more than 50 percent of his genetic material is nonhuman?

Bioethicists and philosophers are watching this borderline closely, asking when it will be breached and who has the authority to guard that border. Philosopher Michael Sandel notes,

> Breakthroughs in genetics present us with a promise and a predicament. The promise is that we may soon be able to treat and prevent a host of debilitating diseases. The predicament is that our newfound genetic knowledge may also enable us to manipulate our own nature . . . to make ourselves "better than well."[9]

Why is this a problem? Sandel worries that the ability to design aspects of our own nature in order to satisfy our own desires would exacerbate our innate drive toward mastery:

> To acknowledge the giftedness of life is to recognize that our talents and powers are not wholly our own doing, despite the effort we expend to develop and exercise them. It is also to recognize that not everything in the world is open to whatever use we may desire or devise. Appreciating the gifted quality of life constrains the Promethean project and conduces to a certain humility. It is in part a religious sensibility. But its resonance reaches beyond religion.[10]

Germline genetic engineering remains out of reach for now. Somatic genetic engineering in humans remains fraught with problems. While genetic testing is growing, the immediate promise of new and far-reaching cures seems to have moved to another arena, that of stem cell research.

Stem Cells and Therapeutic Cloning

When pollsters ask Americans about science and morality, the technology they all seem to know about is stem cell research. We hear a lot about this science on the evening news. The political debate, often shrill, has rocked Congress and the White House and is a wedge issue in political campaigns. Year by year, the public has grown more favorable to stem cell technology. One of the reasons is the beauty of its promise: science hopes to avoid the complications of genetic engineering by instead replacing the entire cell.

Stem cells have two special properties: they are capable of division for long periods of time, and while they are themselves unspecialized, they can be induced to differentiate into other types of cells. There are two types of stem cells: embryonic stem cells and adult stem cells. Embryonic stem cells are those found in a three-to-five-day-old embryo, known as a blastocyst. These cells are the precursors of the entire body. They are pluripotent, meaning they have the capability of differentiating into any of the over two hundred types of cells found in the adult body. Adult stem cells are found in the fully developed body, primarily in the bone marrow but also in blood, muscle, skin, and brain; these cells generate replacements for cells in these tissues that are damaged through injury, disease, or natural wear and tear. Adult stem cells lack the versatility of embryonic stem cells. While some seem to have the potential to differentiate into a number of cell types, they are not pluripotent, and most form cell types specific to the tissue in which they reside. Adult stem cells are also much more short lived in laboratory conditions than embryonic stem cells.

The key to stem cell therapies is to cause the cell to differentiate and propagate under laboratory conditions. If scientists could isolate the genetic cues that cause differentiation, they would have the possibility of developing cell-based therapies for a variety of conditions. Stem cells could be used to provide new neurons for victims of Parkinson's, Alzheimer's, or a stroke. They could provide heart muscle cells for those with congestive heart failure; new nerve cells for paraplegics or those, like Christopher Reeve, with spinal cord damage; skin cells for burn victims; and bone cells for victims of arthritis. New organs could be grown in the lab for transplant.

My father died of Parkinson's disease in 2006, after a thirteen-year struggle with the disease that roughly paralleled that of Pope John Paul II. Like so many others, my family watched the man we loved die by inches. Patients with Parkinson's experience a gradual deterioration of motor control, resulting in loss of balance, difficulty walking, and, for many, debilitating tremors. Ultimately, deterioration of the muscles of the throat and mouth make speech a struggle and then swallowing. Many Parkinson's patients die from choking or of pneumonia exacerbated by the inability to cough and clear the lungs.

Parkinson's is caused by the loss of neurons in the brain that produce dopamine. The goal of stem cell research would be to find a mechanism to replace these cells. Early research has been done using fetal brain cells. These cells are already the right type, which eliminates the need to find the genetic trigger that would cause a basic embryonic stem cell to differentiate into a dopamine-producing neuron. The brain cells are propagated in the lab and then injected, through small holes drilled in the skull, directly into the brain of the recipient, where they set up shop and begin producing the missing dopamine. This process has been used successfully to suppress symptoms of an induced type of Parkinson's in animals, including primates. A similar process, in a human trial, did result in an amelioration of Parkinson's symptoms, but with serious side effects.[11]

The brain has a suppressed immune response, so fetal brain cells, with a genetic pattern dissimilar to the recipient's, could be used in the trial described above. In other parts of the body, such therapies run into the same problem we noted with genetic therapy; cells or organs grown in the lab for transplant would necessarily have a different genetic code from that of the recipient, which would trigger an immune response, just as current organ transplants do. While there are several possible ways to mitigate this problem, the ideal solution would be to match the genetic code of the stem cell to that of the recipient through somatic cell nuclear transfer. In other words, we would produce a blastocyst with the recipient's DNA by starting with an egg whose own DNA has been removed and replaced with that of the recipient. This blastocyst would be a clone of the recipient. Its genetic material would be the recipient's own, thus far less subject to attack from the immune system. In other words, cloning.

This is not the realm of science fiction. Scientists have been cloning animals since the 1950s. In 1996, Ian Wilmut successfully produced the world's most famous sheep, Dolly, the first cloned mammal. Since then, scientists have cloned a number of mammals, including mice, cats, pigs, goats, cows, mules, and rabbits. So far, it has proven difficult to clone primates using the same technique that produced Dolly. It seems that, in primates, removal of the nucleus also removes key proteins needed for proper cell division.[12] The prospect of human cloning has raised valid concerns regarding its safety. Cloned animals tend to be much shorter lived than usual, and many cloned offspring have shown serious genetic abnormalities.

Beyond the technical difficulties, human cloning has also met with resistance on ethical grounds. The idea of reproductive cloning raises visions of carbon copies of ourselves walking around, of Brave New World scenarios of armies of identical workers or soldiers or of parents desperately trying to replace a lost child. Revulsion at such scenarios seems natural. As President Bill Clinton said, at a 1997 press conference in which he called for a ban on human reproductive cloning, "Each human life is unique, born of a miracle that reaches beyond laboratory science. I believe we must respect this profound gift and resist the temptation to replicate ourselves."[13]

In therapeutic cloning, however, we are not talking about creating a new human being. The technology uses stem cells to produce new tissues. Therefore, the entire ethical debate hinges on two matters: the source of the stem cells and the alteration of those cells during the process of nuclear transfer. Early research, including that of James Thomson, the University of Wisconsin biologist who was the first to successfully isolate embryonic stem cells, was conducted on two-day-old embryos donated by infertile couples. Fertility clinics routinely produce more embryos than needed for in vitro fertilization; extra embryos are frozen, and most are discarded once the couple has no further need of them. Supporters of stem cell research point to the fact that, since these embryos are discarded, why not use them to alleviate the suffering of others? Opponents view the use of these cells as the murder of innocent human life. For those who believe human personhood begins at conception, the use, and in that use, destruction, of embryos even a few days in age constitutes the preference of one human life over another. In August 2000, the Vatican issued a statement equating stem cell research with infanticide:

> The living human embryo is—from the moment of the union of the gametes—a human subject with a well defined identity, which from that point begins its own coordinated, continuous and gradual development, such that at no later stage can it be considered as a simple mass of cells. From this it follows that as a human individual it has the right to its own life; and therefore every intervention which is not in favour of the embryo is an act which violates that right.[14]

John Paul's declaration was followed the next day by an address by President George W. Bush in which he announced that no public funds could be used to make new embryonic stem cell lines, effectively cutting off embryonic stem cell research in the United States.[15]

Stem cell research has been particularly contentious in the United States. Theologian Ted Peters rightly points out that much of the energy in this debate originates in the abortion debate. While this debate centers primarily on the question of when human personhood begins and what, therefore, are the attendant rights of the embryo, there are two issues that go beyond this question.

First of all, the use of either embryos or aborted fetuses for research suggests a slippery slope that could lead to the devaluation of all human life. "The devaluation of humans at the very commencement of life encourages a policy of sacrificing the vulnerable that could ultimately put other humans at risk, such as those with disabilities and the aged, through a new eugenics of euthanasia," notes Food and Drug Administration commissioner Frank Young.[16] Peters outlines the second policy concern, which is the force of the marketplace: "This might encourage couples to fertilize ova for the purposes of sale or donation and . . . it might encourage abortions for harvesting [stem] cells."[17] Given the number of blastocysts currently slated for destruction at fertility clinics around the world, this seems a remote eventuality.

Another remote eventuality also grabs our attention—the massive genetic reprogramming of any kind of cell in our bodies. As already noted, adult stem cells generally lack the pluripotency of embryonic cells. They are already programmed to become a particular type of cell.[18] However, it may soon become possible to reprogram any cell to become more like embryonic cells. In 2006, a team of Japanese researchers inserted four mouse embryonic cell genes into an adult stem cell, causing it to behave with some of the versatility of an embryonic cell. In 2008, scientists at UCLA replicated the same process, inducing the pluripotency of embryonic cells in human skin cells. While this avenue of research looks promising in theory, a second look shows that it carries its own moral difficulty, presenting a new quandary for the philosophers and the lawyers. If a cell can be reprogrammed to become pluripotent, is it now a human embryo? Is every cell in our body a potential embryo, waiting only for the necessary genetic reprogramming? While this strikes most of us as absurd, it points to the difficulty of assigning moral status on the basis of potentiality. Each cell in our body contains the genetic potential to become a human person, yet we instinctively know that a cell from our blood, our skin, a particular organ, is not in itself a person.

Our religious traditions have been alert to both the experiments and the speculations. Like the Christian tradition, Muslims have addressed genetic engineering and stem cells, but with slightly different answers. The Islamic distinctions are subtle, as expressed in 1998 by the Islamic Fiqh Academy (a subsidiary of the Organization of the Islamic Conference and World Muslim League). The academy judged the therapeutic use of genetic engineering as valid, provided it did not harm the patient. In considering germline engineering, the members stated the necessity of examining possible harm to future generations. They found alterations that would change the nature of the human body or person to be impermissible. In this, they invoked Surah 4:119, which considers the alteration of the created order as a satanic act: "I will mislead them, and I will create in them false desires; I will order them to slit the ears of cattle, and to deface the (fair) nature created by Allah. Whoever, forsaking Allah, takes Satan for a friend, of a surety suffers a loss that is manifest."[19]

As for stem cell research, Muslim jurists make a clear distinction between the embryo that exists in the first forty days of a pregnancy and later in the term. In the first forty days, the embryo is considered a potential but not an actual human being. This is similar to the understanding found in the Jewish Talmud, which states that an embryo is "mere water" for the first forty days. Dr. Muzammil

Siddiqi, president of the Fiqh Council of North America, also notes that a frozen embryo not implanted in the womb is not in its natural environment and has no chance of becoming a human life without implantation; thus, there is no restriction in using such an embryo for research. But, as with the Christian theologians, Siddiqi recognizes the potential for commercial abuse in producing embryos merely for harvest.[20]

Other Muslim leaders have declared themselves publicly in the face of genetic engineering, issuing *fatwas*, or official religious rulings. Such is the case with Sheikh Yusuf al-Qaradawi. His fatwa permits animal cloning, but to defend human individuality and the necessity of having two parents, he prohibits the cloning of an entire human being. He says the suffering of animals, such as experienced by the famous cloned sheep Dolly, should make us wary of applying these techniques to ourselves. "Genetic engineering has only proven to be a nightmare since animal cloning," the sheikh says. However, he does sanction cloning of human cells to produce new organs or cells for human therapy.[21]

Before the advent of genetic technologies, it was relatively easy to answer the question, "What does it mean to be human?" Today, our technology has led us to see that there may not be a clear answer. We are called to heal the sick and help the suffering, acts for which genetic therapies and stem cell technologies hold tremendous promise. We are also called to respect all human life, not to shed innocent blood, and to protect the least among us. Do those who suffer as Pam Summerhayes suffered, as my father suffered, have a claim on us to do all we can to relieve their suffering? Or is the status of the human embryo as a person such that not even life-saving medical benefits can offset the moral cost of its destruction?

Pharmaceuticals

A diagnosis of HIV infection used to be a death sentence, limiting the patient's remaining life span to ten years or less. The infection makes a person vulnerable to the progressive illness called "acquired immunodeficiency syndrome" or AIDS, which makes one unable to fight threatening microbes in the environment. Today, antiretroviral drugs have been shown to slow the progression of HIV infection, thus delaying the onset of AIDS for up to twenty years. According to the World Health Organization, 39.5 million people worldwide were living with AIDS in 2006. There were 4.3 million new infections in that year, 65 percent of them in sub-Saharan Africa. More than half of the new infections each year occur in young people between the ages of fifteen and twenty-four. Thanks to the new drugs, many of these young people will gain an extra ten years of life, allowing them to be productive members of their economy, raise children, and perpetuate the values and structure of their society.[22]

AIDS is not the only disease plaguing sub-Saharan Africa. Every day three thousand children die of malaria, a parasitic disease for which we now have adequate technologies of prevention and cure. Closer to home, death rates associated with cancer have declined significantly, partly due to the availability of new pharmaceutical treatments. The advent of the polio vaccine, the worldwide eradication of smallpox, and treatments for typhoid, cholera, and tuberculosis all show how pharmaceutical technology provides a vital service in alleviating human suffering. Which of us would want to be without modern antibiotics or analgesics? Pharmaceuticals save lives and could save even more were we to find a way to broaden access to them for the poor and those in developing countries.

Like the technologies we have already seen, pharmacology has given us tremendous therapeutic results. However, like the previous technologies of the body, pharmacology also raises the prospect of human enhancement. It gives us the ability to change the human organism and raises the question of what is dysfunctional and what is normal. Where is the line between aiding the sick and suffering and changing the human condition? This question is particularly acute in the rapidly expanding market of neuropharmaceuticals.

Prozac: The Personality on Steroids

According to the U.S. Centers for Disease Control, antidepressants were the most frequently prescribed class of drugs in 2005, with a total of 118 million prescriptions. The worldwide market for Prozac (fluoxetine), Zoloft (sertraline), Paxil (paroxetine), and others totals more than $10 billion a year. These drugs are selective serotonin reuptake inhibitors, or SSRIs. They block the reabsorption of the neurotransmitter serotonin, thus increasing the amount of serotonin available to the brain. Serotonin is one of the primary chemicals that the brain uses to transfer signals between neurons and plays a major role in regulating mood, appetite, anger, sleep, sexuality, and body temperature. Low levels of serotonin are associated with depression, anxiety, aggression, compulsion, and suicide. SSRIs are used today not only in the treatment of depression but also of anxiety, panic disorders, obsessive-compulsive disorders, eating disorders, posttraumatic stress disorder, Tourette syndrome, irritable bowl syndrome, and chronic pain.

The advent of SSRIs and other psychotropic drugs reflects the rapid advances that have been made in understanding the physical and biochemical structure of the brain. We now know, for example, that persons suffering from depression show decreased levels not only of serotonin but also of another neurotransmitter, norepinephrine. They also show increased levels throughout the day of the stress hormone cortisol. While depression has long been associated with troubling thoughts and emotions, or even with a certain personality type, we now know that it is truly a disease, associated with nerve-cell atrophy in the brain that is cumulative over time. Depression also harms the heart, endocrine system, and skeletal structure, and can be life-threatening since it can lead to suicide.[23] The advent of new medications to treat this illness is clearly a step forward.

Yet many people remain troubled. Psychiatrist Peter Kramer notes that, after the publication of his book *Listening to Prozac*, in which he describes the almost miraculous transformation of many of his patients after taking Prozac, he is frequently asked the question, "What if Prozac had been available in van Gogh's time?" The unease that underlies this question is twofold. First, might widespread use of such a drug deprive us of some of the world's most insightful art, music, or literature? Would its use remove an aspect of the human condition? Second, Prozac and other psychotropic drugs have the potential to radically alter a patient's personality. This then raises the question of where personality comes from. Is our personality strictly biochemical and alterable at will? Who are we, then?

In an article in *Psychiatric Times*, psychiatrist Kevin Kelly describes a fifty-five-year-old man who came to him feeling dissatisfaction with his work. The man had difficulty making decisions, a difficulty that seemed to be rooted in an aversion to expressing aggression. Dr. Kelly found that his patient became comfortable expressing differing levels of aggression based on the dosage of Prozac given. Simply put, the more Prozac, the more aggressive the patient became. Dr. Kelly notes, "The medication had allowed us to establish a pharmacotechnology of aggression, placing him anywhere along a spectrum from timid to abrasive, simply by titrating the dose. But the question of where he wanted to be on this spectrum, or where he 'should' be, or where the optimally healthy position was located, could not be answered by pharmacology."[24]

Kramer tells a similar story. He describes a patient, Tess, who becomes significantly more self-confident, assertive, relaxed, and energetic on Prozac. Within weeks, she changes her circle of friends, ends a dissatisfying relationship, and changes her management style at work. Tess eventually goes off the medication, but after eight months, as she finds herself becoming progressively inhibited and subdued, asks to go back on it, saying, "I'm not myself." Kramer asks, "But who had she been all those years if not herself? Had medication somehow removed a false self and replaced it with a true one? Might Tess, absent the invention of the modern antidepressant, have lived her whole life—a successful life, perhaps, by external standards—and never been herself?"[25]

Both these stories raise the question of authenticity, of where a person's true personality resides. If a child who has suffered from untreated asthma were to become more self-confident, assertive, and boisterous on having his asthma treated, we would find this a normal response and say that his true personality was finally able to flourish. When the illness is mental, such a response becomes a bit more problematic, though severe clinical depression does seem to strongly inhibit self-expression.[26] But what about Dr. Kelly's patient, who did not suffer from clinical depression but whose personality changed on Prozac? We have now moved from the realm of therapy to that of enhancement. To what extent should one be able to craft one's own personality through drug use? How does this differ from the kind of recreational drug use society has traditionally condemned?

Prozac functions in part by blunting one's emotional response to the world. When that response is a painful one, Prozac can provide relief. But does this relief come too easily? Irish psychiatrist Maurice Drury, in an essay titled "Madness and Religion," describes a Catholic priest for whom saying the mass no longer held any meaning. This man felt he had lost his faith, was having trouble sleeping and eating, and was experiencing odd physical pains. While the latter are the classic symptoms of clinical depression, the priest found meaning for his state in the former; he attributed his malaise to the realm of the spirit.[27] The struggle with the dark side of the soul is a recurring theme in religious literature. We need only think of St. John of the Cross' "dark night of the soul." Or consider the psalmist, crying out:

> How long, O Lord? Will you forget me forever?
> How long will you hide your face from me?
> How long must I bear pain in my soul,
> And have sorrow in my heart all day long? (Psalm 13)

Russian novelist Leo Tolstoy writes of a similar time of spiritual darkness: "I can call this by no other name than that of a thirst for

God. This craving for God had nothing to do with the movement of my ideas,—in fact, it was the direct contrary of that movement, but it came from my heart. It was like a feeling of dread that made me seem like an orphan and isolated in the midst of all these things that were so foreign."[28] It was this search, this time of despair that led Tolstoy to change his entire way of life. Gordon Allport notes that "the mature religious sentiment is ordinarily fashioned in the workshop of doubt."[29] The maturing of the individual necessitates some pain.

However, the level of emotional pain many experience seems to be greater than in the past. Depression is ten times more common among those born in the middle third of the last century than those born in the first third. In 1957, depression was a rare diagnosis, affecting fewer than fifty people per million. It is now estimated that it affects up to 121 million worldwide.[30] Twice as many women as men are diagnosed with clinical depression.[31] This statistic suggests to some that the illness is less an individual ailment than an indictment of our society.

That is indeed the indictment made by some feminist authors. They argue that female depression is an appropriate response to a patriarchal system. Medicating women can be viewed as one aspect of societal forces that seek to push all of us toward a personality type that is outgoing, vivacious, assertive, energetic, pleasure oriented, and uninhibited, a cultural norm that has traditionally been associated with masculinity. As one feminist blogger writes, "What sort of culture do we become when we can gender engineer ourselves right into the sort of personality types that kick ass in business, that make us less sentimental about sex, and less overly sensitive to the needs of others?"[32]

These writers see the widespread use of Prozac among women as a reinforcement of the social, political, and economic policies that have devalued both the roles of women and the psychological strengths traditionally associated with the feminine. Prozac can be a substitute for social reform, hiding the alienation women might normally feel in a society that rewards the masculine. The philosopher Jacques Ellul has observed how the growth of technology in general imposes a lock-step to social life and blinds us to that life's contours:

> When a society becomes increasingly totalitarian . . . [it] requires its citizens to be conformist in the same degree. Thus, technique becomes all the more necessary. I have no doubt that it makes men better balanced and "happier." And there is the danger. It makes men happy in a milieu which normally would have made them unhappy.[33]

Ritalin: The Brain on Steroids

While feminists see the increasing use of Prozac as a way to masculinize women to better fit a capitalist society, young men who exhibit too much energy or assertiveness or too little inhibition have their own psychotropic drug. Ritalin, the trade name for methylphenidate, is a stimulant in the same family as methamphetamine and cocaine. Ritalin is prescribed to over 3.5 million children and adults around the world (80 percent of whom are in the United States) to treat the syndrome known variously as attention deficit disorder (ADD) or attention deficit-hyperactivity disorder (ADHD). ADD was first listed as a disease in the American Psychiatric Association's *Diagnostic and Statistical Manual of Mental Disorders* in 1980, though Ritalin had been used experimentally since the early 1970s (the listing has since been changed to ADHD). ADHD is diagnosed through its symptoms, which include restlessness, hyperactivity, difficulty concentrating, and impulsive behaviors. The striking thing is how rapidly the number of diagnoses of ADHD has risen. In 1990, 900,000 children were using Ritalin. By 1997, it was 2 million. By 2000, 3.5 million were using Ritalin, with another 1.4 million using other related stimulant drugs.[34] Some researchers suggest that up to 15 million Americans may exhibit some form of ADHD, while others note that symptoms can be seen in anywhere from 7 to 17 percent of school-age children, with up to 20 percent of the boys in some school systems receiving stimulant medication.[35]

Why such a rapid rise? Some authors suggest that ADHD is not really a disease at all but a diagnosis for a collection of normal childhood behaviors.[36] There is no clinical test for ADHD. Diagnosis is based on a collection of behaviors and often not finalized until beneficial effects are observed. Physician Lawrence Diller writes:

> We are left with the possibility that ADD may be a catch-all condition encompassing a variety of children's behavioral problems with various causes, both biologically predetermined and psychosocial. And the fact that Ritalin helps with so many problems may be encouraging the ADD diagnosis to expand its boundaries.[37]

Others point to the speed and availability of sensory stimulation in our culture as the source of hyperactivity among our children. Children whose minds are conditioned by the rapid-fire world of MTV or most video games find themselves uncomfortable in the absence of such stimuli.[38] Recent studies suggest that there is a genetic factor that predisposes one to ADHD, but most researchers agree that the environment also plays a role.[39]

Whatever the source and whether the diagnosis is, in most cases, accurate, what we see in this diagnosis is a move toward a biological understanding of human behavior. The President's Council on Bioethics describes the problem inherent in this medicalization of our understanding:

> A medical diagnosis of ADHD . . . implies or claims the presence of a malfunction in the child's brain. If impulse control is the behavioral product of combining an impulse-to-do and the will-to-restrain, one can then imagine Ritalin as acting to reduce impulse-to-do rather than strengthen the will-to-restrain. In contrast, the

> traditional tools of teaching young children "good" and "bad" behavior, involves praise and blame from parents and teachers. . . . A central moral question about treating hyperactive children with Ritalin is now apparent: Is it desirable to substitute the language and methods of medicine for the language and methods of morals?[40]

Advances in neuroscience have led to a general move in psychiatry away from its Freudian roots, which understood our mental world to be formed through the social milieus of environment and experience, toward a view that locates the source of mental difficulty in the individual's brain chemistry. This view lends itself nicely to a technological fix. Change the chemistry in the brain with the appropriate drug and the problem is solved. Or is it? Just as taking an aspirin might alleviate a headache without getting at the initial cause of the headache, so taking Ritalin or Prozac changes an individual's behavior or emotions without addressing the cause of that behavior or those emotions.

Just as Prozac might be used by those who are not diagnosed as clinically depressed as a tool for personality enhancement, Ritalin is increasingly used by adults or college-age students to enhance mental performance. According to the *New York Times*, recent campus surveys show that as many as 20 percent of college students have at one time used Ritalin as a study aid or as a boost to concentration for writing a paper or taking an exam.[41] As with the question of where a person's true personality lies, Ritalin use raises the question of what is a true measure of one's mental abilities, those measured on or off Ritalin.

Christians and Jews find little in their scriptures beyond the injunction to heal the sick and help the suffering that would aid in deciding whether to take psychotropic drugs. However, the use of Ritalin and Prozac raises an interesting question for Muslims. In general, substances that cloud the mind are forbidden, or *haram*:

> O ye who believe, intoxicants, gambling, sacrificing to idols and divining by arrows are but abominations and Satanic devices. So turn wholly away from each of them that you may prosper. Satan desires only to create enmity and hatred between you by means of intoxicants and gambling and to keep you back from the remembrance of Allah and from prayer. (5:90–91).

Muslims generally consider all nonprescribed drugs as dangerous to both the body and to one's faithfulness. This would suggest that using Ritalin as a study aid would be frowned upon, while it would be acceptable when prescribed for a child by a doctor. Yet neuropharmaceuticals can clearly be abused, and since they do alter one's mental perception, they would seem to require a certain level of scrutiny.

Performance as Our Standard

As I write, in the summer of 2007, the newspapers are filled with two sports stories. The first is the ongoing Tour de France. So far the prerace favorite and the racer wearing the yellow jersey (holding the lead) on day sixteen have been disqualified for doping. Olympic officials are threatening to eject cycling as an Olympic sport due to the recent prevalence of drug-related scandals. Meanwhile, in the world of baseball, Barry Bonds grimly chases Hank Aaron's record while facing similar accusations of steroid use.

Steroids make sports no longer a contest but an exhibition. The idea of contest implies a somewhat level playing field, in which those who excel do so through training and perseverance, albeit with certain gifts of nature. The assumption has been that, while some might be born with a body better suited to a certain sport, the differences from person to person were limited. Doping dissolves the limits to human performance that have long defined what a normal person can do. The question is whether this matters. Some think not, so long as the option is available to all. Professional sports are "entertainment, business, pure and simple," according to David Malloy, who teaches sports ethics at the University of Regina. "I don't think the average person cares that players are on steroids. I think they just want to see them hit the ball a mile out of the park."[42]

Malloy's remarks highlight the dilemma of using pharmaceuticals for either mental or physical enhancement. Both demonstrate an obsession with performance that is characteristic of our technological society. That performance should become the standard by which we measure ourselves or are measured by others is no surprise in a society dominated by technique. Technology itself is measured in terms of concrete results, and this measure is generally quantitative. In a technological society, we measure human performance in the same way, the quantitative results of higher batting averages, faster race times, or higher test scores. Even Peter Kramer's patients on Prozac partially evaluated their happiness in terms of the number of dates they were asked for or a raise in salary at work.

This obsession with performance shows a shift from understanding ourselves in terms of who we are to understanding ourselves in terms of what we do. And doing is a never-ending task; there is no point at which we will not feel the need to do more, to break the record one more time, to achieve an ever higher standard. Jacques Ellul notes the end result:

> There is no longer respite for reflecting or choosing or adapting oneself, or for acting or wishing or pulling oneself together. The rule of life is: No sooner said than done. Life has become a racecourse . . . a succession of objective events which drag us along and lead us astray without anything affording us the possibility of standing apart, taking stock, and ceasing to act.[43]

The growing use of pharmaceuticals to enhance our personalities, minds, and bodies places us on a treadmill, a never-ending search for greater happiness, better performance, higher standards of strength and beauty. On such a treadmill, one can never be satisfied with oneself "as is." And that will only lead to the sad state of dissatisfaction and alienation that is all too common in our society today. Pass the Prozac.

Bionic Men and Women

The Six-Million-Dollar Man, The Bionic Woman, Robocop, The Terminator—we all know these blockbuster movies and television shows. Their heroes are the cyborgs of science fiction, individuals who are both human and machine. Cyborgs are no longer confined to the realm of science fiction, however, given the advancement of mechanical aids to the human body. The day of a human/machine hybrid is not far in the future, and it will radically influence our concept of what it means to be human. Mechanical enhancement of the human body is already with us; anyone who wears glasses or contact lenses is already enhanced technologically. The implantation of mechanical devices in the human body, such as prosthetic limbs, pacemakers, and hearing aids, carries this aid a step further. Implantation aside, many people depend on therapeutic equipment such as kidney dialysis machines, ventilators, and feeding tubes. All these technologies have an ethical dimension. They force us to ask how far we will go to enhance ourselves with technology, how dependent we wish to be on machines, and how society will apportion the money for people to use these new therapies.

Mechanical Prostheses

The idea of implanting mechanical parts in the human body in order to restore or improve function is not new. Most of us are familiar, and comfortable, with the use of prostheses to replace missing limbs, pacemakers to smooth an erratic heartbeat, artificial knees and hips, lens implants following cataract surgery, or cochlear implants that restore or augment hearing. While early mechanical implants were relatively inert, many today, including the pacemaker and the cochlear implant, involve an intricate combination of biology, mechanics, and electronics. The development of devices of this type is known as biomechatronics and is a growing research field.[44] There are current biomechatronic prostheses to aid motor control, hearing, and vision. These devices connect directly to the physiological and/or neurological systems of the human body.

As an example of where this technology is going, consider prosthetic hands currently under development. In 2006, scientists at the Applied Physics Laboratory at Johns Hopkins were the first to successfully harness signals from the brain to move individual fingers in a mechanical hand. Sensors in the fingers detect heat and cold. Flexible force sensors detect how much pressure to apply when picking up an object or squeezing a ketchup bottle. Other sensors report hand position. Some of these sensors communicate through the patient's skin, while others are implanted in the muscle of the remaining arm. The hope is eventually to develop sensor technology that can be wired directly to the central nervous system.[45] Such prostheses represent a great advance over current mechanical arms and hands, which can grasp objects in response to muscle impulses communicated to sensors on the skin but which are bulky and have a limited range of movement.

Other biomechatronic devices are implanted completely within the user's body. One example of such a system is the cochlear implant, a mechanical device that transforms sound waves into electrical impulses sent to the cochlea in the inner ear to stimulate the auditory nerves. The implant has both external and internal components. An external microphone picks up the sound and transmits it to a processor. The processor converts the sound waves to electronic impulses that are transmitted to a receiver implanted near the auditory nerves of the ear. This signal is detected by the auditory nerve and sent to the brain. Over one 100,000 patients worldwide use such an implant to restore lost hearing.[46] Another 250 patients have had loss of hearing in both ears restored through the use of auditory brainstem implants.[47] Other devices, such as pacemakers and cerebellar stimulators, are completely internal to the body. Researchers at the University of Pennsylvania, led by Kwabena Boahen, have developed retinal implants that detect motion, such as someone shaking his or her head.[48] Electrodes implanted in the brain limit tremors in patients with Parkinson's, mitigate symptoms in Tourette syndrome, and aid in the treatment of depression.[49] All these devices, whether implanted in a sensory organ or in the brain itself, send and receive signals in concert with the brain.

To help people move impaired limbs, we also have technology worn externally, linked to a variety of braces. These can help people with motor control after a stroke or with cerebral palsy or incomplete paraplegia.[50] One example, the Active Ankle-Foot Orthosis (AAFO), developed at MIT, fits over the ankle and foot of a patient and restores movement to a paralyzed or partially paralyzed ankle. A similar exoskeletal apparatus, the Berkeley Lower Extremity Exoskeleton (BLEEX) fits over a wearer's legs and is used to aid in the bearing of weight while walking.[51] Funded by the Defense Advanced Research Project Agency, Berkeley researchers expect the device to have applications in the military, firefighting, and rescue operations.[52] BLEEX moves biomechatronics from the realm of therapy to that of enhancement. It gives the wearer a level of functioning beyond the body's normal capability. The mule and the automobile, of course, might be thought of as mechanical technologies that enhanced our ability to carry weight. BLEEX is new in that it is an enhancement directly attached to the user's body.

Is there a BLEEX equivalent, i.e., a prosthesis that not only restores normal functioning but enhances human performance, on the horizon for our brains? If implanted chips restore hearing or vision, why not use them to enhance sensory capabilities, such as to provide infrared vision or to bestow new cognitive capabilities, such as embedding a calculator in the mind? This is perfect material for science fiction but very risky in real biology, according to Steffan Rosahl of the Department of Neurosurgery at the University of Freiburg. Any implant that exchanges energy and information has the potential to change its surrounding biological system. In other words, Rosahl points out, "The implanted device may interfere with the organ it is implanted into. In the case of sensory implants, the organ of interference is the human brain."[53] A brace that enhances the strength of the leg does not change that leg, but a calculator chip implanted in the brain will surely change the functioning of the mind—and thus the "self" of a person.

The public seems to know this chilling implication. Rosahl notes that it is already possible with current technology to "equip or 'arm' human beings for better performance, better mood or other advantages by means of an electronic implant."[54] Yet, few have lined up for such enhancements.[55] One reason is that sensory implants are permanent. While the army might like the idea of equipping its soldiers with night vision, this feature would not go away on a soldier's decommissioning. Second, the soldier himself is equally unlikely to want to risk permanent loss of vision should such an implantation go awry. Most of us fear the potential loss of a sense of self or loss of control that a cognitive implant could engender. Such an implant could be controlled or at minimum could malfunction in the presence of microwaves, radio waves, or altered magnetic fields.

Nevertheless, there is one arena where we will take the risks—the competitive and glamorous world of sports. Numerous athletes, from professionals down to high-school students, are willing to risk the side effects of performance-enhancing drugs such as steroids or human growth hormone. As yet, mechanical implants are not an issue. But they could be in the near future. Oscar Pistorius of South Africa was ruled eligible to be the first amputee runner to compete in the 2008 Olympics on artificial legs, though he did not make the South African team. Pistorius walks on two J-shaped blades made of carbon fiber. He is an excellent runner who came very close to meeting the Olympic qualifying standards. The question arises; is Pistorius disabled or "too-abled"—in other words, are his prostheses strictly therapeutic, for therapeutic they surely are, or are they also enhancements, a technology that gives him an unfair advantage over regular runners?

The world governing body of track and field has long prohibited technological aids such as wheels or springs. Yet they have allowed high-tech shoes, such as those with air cushions that clearly give added spring to the runner's step. Prostheses such as those worn by Pistorius could give an advantage both in terms of spring and by lengthening his leg, and thus his stride. A question posed on the web site of the Institute for Ethics and Emerging Technologies asks: "Given the arms race nature of competition, will technological advantages cause athletes to do something as seemingly radical as having their healthy natural limbs replaced by artificial ones?"[56]

Technology and End-of-Life Care

Far more people end up connected to machinery not in the form of prostheses or implants but in a medical or hospital setting.[57] Few ethical or religious issues arise when this connection is temporary. No one questions the benefits of diagnostic machinery or the temporary use of mechanical ventilators, heart pumps, or kidney dialysis. The technology can become an issue when the use of the machinery is a permanent necessity for the continuation of the patient's life. This is especially problematic when the patient is not conscious and, therefore, unable to express his or her wishes regarding mechanical assistance.

In many cases, even the permanent use of mechanical technology is an unquestionable good. The actor Christopher Reeve lived a productive life as a speaker, fundraiser, and film director for nine years while connected to a ventilator after suffering paralysis from an injury sustained in an equestrian competition. However, not all cases are as simple. The question of when to use or to discontinue the use of a medical technology rose precipitously in public awareness in 2005, due to publicity surrounding the case of Terri Schiavo. In 1990, Schiavo collapsed with acute respiratory and cardiac arrest, resulting in severe brain damage. She was later diagnosed as being in a permanent vegetative state and was dependent on a feeding tube for nutrition and hydration. After seven years, her husband petitioned for removal of the feeding tube, a petition contested by her parents who believed Schiavo retained elements of consciousness. Her case rose to the level of the Supreme Court, which ruled in favor of the tube's removal. Schiavo died in March 2005. Some critics saw her death as a long-overdue acceptance of what her husband claimed was a previously expressed desire not to be kept alive by artificial means; others claimed it was a case of juridical euthanasia.

The withdrawal of life-support technology is one of the most difficult decisions faced by doctors, patients, and their families. Most wish for their loved ones a peaceful and dignified death, yet many worry that their decision to terminate life support might make them instrumental in that death. The use of such technology is often initiated by emergency-care providers with little or no input from the patient or family. When the family arrives, the emotional stakes can be high. According to ethicists, the decision to end life support can be just as moral as the first decision on whether to use or refuse a life-support technology. Still, the decision to end support is usually more wrenching. Ethicists want to make clear, however, that such strong emotions should not cloud this fact—discontinuing life support is not euthanasia because it does not introduce a new cause of death. Morally speaking, the intention is not to cause death but to ease the physical or psychological burden on either the patient or the patient's family.[58] Yet death usually follows. How are families to make such a decision?

In general, medical professionals use four criteria to determine the legitimacy of discontinuing mechanical life support. First is the presence of a fatal condition. If the patient stands a good chance of recovery, life support should by all means be used. Although we tend to focus on patients such as Terri Schiavo, most medical technologies are used by nonterminal patients—those recovering from surgery, premature infants, and persons recovering from trauma. Among those suffering a fatal condition, the next criterion is the autonomy of the patient. If the patient is conscious, able to communicate, and capable of rational decision making, his or her desires are paramount. A patient has every right to refuse a particular medical treatment.[59] This right to autonomy underlines the importance of patients' leaving an advance directive, in those states and countries where such a directive is acknowledged.

Third, one must ask whether the therapy is effective.[60] If a therapy is futile—in other words, it delivers no medical benefit—it should be stopped. In the case of patients in a permanent vegetative state (PVS), the effectiveness of feeding tubes is not immediately apparent. Feeding tubes are effective in the sense that they have one of the highest success rates of all medical technologies. On the other hand, they in no way address the underlying condition. A feeding tube delivers nutrition and fluids. It cannot improve the patient's underlying health problem and often seems to leave that person suspended between life and death.

Under current laws, doctors may withhold or discontinue treatments that are deemed to be futile. The criteria for futility are that the patient suffers a terminal condition, that the condition is irreversible, and that death is imminent. These criteria are not always clear in the case of PVS patients. For an anencephalous child, one born without a functioning brain, it is clear that the condition is terminal and irreversible and that death will come almost immediately following cessation of mechanical intervention. The same is also true for PVS patients who suffer a degenerative disease such as Alzheimer's or Parkinson's. However, those who develop PVS as a result of trauma or lack of oxygen sometimes exhibit a spontaneous recovery, though these are exceedingly rare after the first twelve months.[61]

The final criterion is whether a given medical treatment places an excessive burden on the patient, family, or community. Determining what constitutes an excessive burden is equally problematic. One patient undergoing indefinite kidney dialysis might consider the prolonged treatment an excessive burden on emotional grounds. Others would not. Prolonged treatment can place much stress on members of the patient's family, particularly when the patient is a child. An adherent to Christian Science or a Jehovah's Witness might consider certain treatments religiously forbidden or undesirable and thus a spiritual burden. And, of course, prolonged treatment is often a financial burden for the patient and/or the family or community.[62]

The question of financial burden raises issues of justice and accessibility. Do the poor deserve life support? In December 2005, Tirhas Habtegiris, a twenty-seven-year-old immigrant from East Africa who was dying of cancer, was disconnected from a ventilator after doctors at Baylor Medical Center determined that further treatment was futile. While it was clear that Habtegiris' condition was terminal and irreversible and that death was imminent, she had hoped to stay alive long enough for her mother to arrive from Africa. The hospital, however, was under no legal obligation to continue care for more than ten days after it had provided notice that such treatment was medically futile. Ms. Habtegiris' mother did not arrive in time.[63] The question of allocating expensive medical resources to the terminally ill or persistently vegetative patient is a vexing one for countries that have state-sponsored medical care, which face the unenviable choice between rationing care in such a way that it hastens the death of some patients or allocating increasing resources in the face of minimal benefits.

From a religious perspective, the decision to use or withhold a given medical technology comes down to two issues. First, is the technology ordinary or extraordinary care? Christians are expected to use all ordinary means to care for the sick and help the suffering. The Catholic Church, however, has a five-hundred-year tradition of permitting patients or their families to forgo extraordinary care. The use of a machine, by its very nature, seems to make such care extraordinary. Yet we see nothing extraordinary these days in the use of machines for diagnostics (X-rays, CAT scans, etc.) or the use of an ambulance to transport a patient to the hospital. Each of these machines is used temporarily. The permanent need for life-sustaining machinery, such as a ventilator or dialysis machine, in the face of terminal illness is more extraordinary. Such machinery need not be considered a part of ordinary care.

Feeding tubes are considered a special case. The Vatican recently ruled that feeding tubes should be considered "in principle, an ordinary means of preserving life."[64] This does not, however, mean that feeding tubes are morally obligatory in all cases. Pope John Paul II, in the late stages of Parkinson's, refused to have such a tube implanted, and his subsequent death was in no way considered to be suicide. The deciding factor here was the fact that Parkinson's is a progressive illness; John Paul's death was imminent, with or without a feeding tube. The Vatican document also allows for the withdrawal of tube feeding if it is futile or if the patient is experiencing physical suffering, either as a result of the tube itself or as a result of some concurrent and untreatable condition.[65]

Islam agrees that human life is sacred, not to be in any way deliberately shortened. The Qur'an enjoins: "Take not life, which Allah has made sacred, except by way of justice and law" (6:151), and Surah 5:32 states that to take one human life is as if one has killed all humankind. It also notes (39:10) that one should be patient with pain and suffering. The Islamic Medical Association of North America, like the Vatican, has noted that if treatment is deemed futile, it can be discontinued, particularly in cases of brain death or a persistent vegetative state. However, the basic human rights of hydration, nutrition, nursing, and pain relief must be given. A feeding tube is not considered extraordinary care, yet it can be withdrawn if deemed futile and should not be reinserted.[66]

Christians, Jews, and Muslims base their arguments on the sanctity of life. Must life be sustained at all costs? The terminally ill present us with the example of the most helpless in society and as such demand our protection of their dignity and human rights. However, that dignity includes the right to a dignified and peaceful death. If prolonging life mechanically is of no benefit to the patient, in what way does it add to his or her dignity?

To conclude, this understanding of sanctity returns us to Genesis and the concept of humans created in the image of God. As we saw earlier, the book of Genesis does not spell out this image. Theologians such as Karl Barth, however, remind Christians that the image is about relationships, based on the three persons of the Trinity. Hence, relationship to God and others, not physical life itself, is the ultimate good. A person's intrinsic value and relationship to God and to others does not end at death. In our daily lives, moreover, we relate to people not in the abstract but very concretely, for as the theologian Stanley Hauerwas points out, "We care or do not care for them because they are Uncle Charlie, or my father, or a good friend."[67] That relationship also extends beyond death in the Christian idea of the communion of saints, a realm in which the living and the "dead" pray for each other. Thus, death does not end human-to-human relationships, any more than it ends our relationship to God. Therefore, life itself cannot be the ultimate good. This is also true in Islam, where submission to God is the ultimate good, even when that submission may, in fact, lead to death.

The contrast between these theological definitions of our humanity and the cold technology that surrounds us can be stark. What has the communion of saints got to do with Prozac, a genetic alteration, or a technological implant in the body? Each of these technologies raises the question of what it means to be human, of when suffering is a normal part of the human condition and when it is an exceptional thing that we must do everything in our power to alleviate. They ask us, in an unprecedented way, to make decisions for ourselves and our loved ones as to when life should be sustained and by what means. That these technologies benefit our community is obvious. They present new possibilities for us of greater ability, greater happiness, and longer life. The cure of various afflictions of body and mind strengthen not only the individual but the whole community. Yet we do well to be wary of the use of these technologies for enhancement when such enhancement could be divisive of the human community. Thus, we falter on the third question raised by the Amish, which asks whether a technology changes the nature of the community itself. Genetic engineering, neuropharmaceuticals, and the mechanical equipment of health care can heal, can divide, and can become an excessive burden, something we need to forgo. Knowing which of these situations is the case is often extremely difficult and requires the best wisdom of the community, present and past, to aid those faced with choosing or denying the use of medical technologies.

CHAPTER 3
Cyberspace on Our Minds

TWO DIFFERENT PROFESSIONS, the computer scientist and the brain researcher, are rallying around a single metaphor for our intellectual lives. Our physical brain is like the "hardware" of a computer, they say, while our mind and mental activity are akin to "software." This shows just how much computer technology has shaped our modern understanding of human nature. Technology is literally changing our minds. In the last chapter, we saw how drugs can do this by altering the hardware of our brains. The virtual world of computers alters both the hardware and the software. Computers have changed how we define intelligence and what we call reality in our social and aesthetic lives. These machines may be altering the very biology of our brains as well.

Computers give us a means to amplify or enhance our mental experience. We seek an alternative to the corporeal mind in artificial intelligence (AI), we use computers as tools to expand our mental powers, and, increasingly, we use them to provide a new world for us to explore and in which we can interact with one another. We speak of cyberspace as a place, a place that some young people find increasingly hospitable.

The computer revolution has also deeply engaged the world of religion. Despite the assertion by some that religion is "otherworldly," it is, in fact, fundamentally anchored to this physical world. Any altered perceptions of the world will very likely influence religion. In the Gospel of John, Jesus asks his followers to plunge into the world: "I am not asking you to take them out of the world, but I ask you to protect them from the evil one" (John 17:15). The church itself remains in the physical world, declared Pope Paul VI. "She remains as a sign—simultaneously obscure and luminous—of a new presence of Jesus, of His departure and of His permanent presence."[1] Muhammad also saw life as being deeply engaged in the world. He eschewed monasticism, expecting his followers to live active lives rather than only "sit in the corner of a mosque in expectation of prayers."[2] Now, computers are changing that world. They are forcing us to assess how the "virtual" world may affect our "real"-world relations to God, others, and the natural order.

Human and Artificial Intelligence

What is the mind? Where is consciousness? Do we have souls, and if so, how do they relate to the mind or the body? These questions underlie many of the ethical issues that bedevil both scientists and politicians in the twenty-first century. Controversies surrounding abortion and stem-cell research are rooted in varying views of when a new human soul comes into being. Euthanasia asks a similar question in reverse, namely, when does the soul depart in death? If our mental reality is, as the computer scientists and brain researchers suggest, analogous to the hardware and software of a computer, we must now also ask whether a computer itself could have a soul.

The key to this puzzle is in how we define *intelligence* and *soul*. These two concepts are not as easy to define as we might expect. Most of us have an intuitive understanding of what we and others mean when using those terms, but would find it difficult to give a satisfactory definition of either. Yet a clearer understanding of both is crucial if we are to sort through the controversies mentioned above. Our approaches to designing artificially intelligent machines give us one avenue toward understanding the different aspects of intelligence and how they relate to the vexing question of the soul.

Three Approaches to Intelligence

In designing an AI computer, the first approach has been to define intelligence as the ability to solve problems. This definition fits our intuitive notion of intelligence, based as it is on the model of activities that we consider indicative of highly intelligent people, such as the ability to play chess or solve complicated equations in mathematics or physics. Computer scientists in the 1960s and 1970s took mathematics as a model (after all, most early computer scientists were mathematicians). Just as the field of geometry is built from a finite set of axioms and primitive objects such as points and lines, so early AI researchers, following rationalist philosophers such as Ludwig Wittgenstein and Alfred North Whitehead, presumed that human thought could be represented as a set of basic facts. These facts would then be combined, according to set rules, into more complex ideas. This approach to AI has been called "symbolic AI." It assumes thinking is basically an internal process of symbol manipulation.

Symbolic AI met with immediate success in areas where problems can be described using a limited set of objects or concepts that operate in a highly rule-based manner. Game playing is an obvious example. The game of chess takes place in a world in which the only objects are the thirty-two pieces moving on a eight-by-eight board, and these objects are moved according to a limited number of rules. Other successes for symbolic AI occurred rapidly in similarly restricted domains, such as chemical analysis, medical diagnosis, and mathematics. These early successes led to a number of remarkably optimistic predictions of the prospects for symbolic AI.

Symbolic AI faltered, however, not on difficult problems like passing a calculus exam but on the easy things a child can do, such as recognizing a face in various settings, riding a bicycle, or understanding a simple story. One problem with symbolic programs is that they tend to break down at the edges; in other words, they cannot function outside or near the edge of their domain of expertise since they lack knowledge outside of that domain, knowledge that we think of as common sense.[3] Humans make use of millions of bits of knowledge, both consciously and subconsciously. Often we do not know what bits of knowledge or intuition we have brought to bear on a problem in our subconscious minds. Symbolic AI programs also lack flexibility. In 1997, the computer Deep Blue beat then reigning world chess champion Gary Kasparov. In the ten years since, Deep Blue's successors continue to play chess—and only chess. Kasparov, on the other hand, has become a politician and presidential candidate in Russia. Intelligence seems to be a quality that is larger than can be captured in any symbolic system.

A second way to look at intelligence is as the ability to act within an environment. This means that, first of all, intelligence is embodied. Of course, any intelligent agent would be embodied in some way. Deep Blue did not have what we would think of as a body; it could not pick up the chess pieces and physically move them. However, the program was embodied in a bank of supercomputers. So the question is not whether intelligence requires a physical body but what kind of body. Does a human-like intelligence require a human-like body?

Our bodies determine much of the nature of our interaction with the world around us. Our perception is limited by our physical abilities. For example, we think of location in two-dimensional terms because we walk rather than fly. We evaluate a situation primarily by sight and sound; anyone who has ever walked a dog knows that the animal evaluates with nose to the ground, receiving a whole different data set from what we see. In a similar fashion, our physical makeup determines how we act within the world. Our opposable thumbs, combined with the softness and pliability of our skin and underlying muscles, allow us to grasp and manipulate objects easily. When we ride a bicycle, we need not calculate equations of force, trajectory, and balance; our muscles, nerves, and inner ear do the work for us. In fact, should we begin to make such calculations consciously, we are likely to fall off! Most athletes know that they perform at their best when their minds are in a meditative, rather than a discursive, mode. As one of the characters says in Frederich Nietzsche's *Thus Spake Zarathustra,* "Behind your thoughts and feelings, my brother, there stands a mighty ruler, an unknown sage—whose name is self. In your body he dwells; he is your body. There is more reason in your body than in your best wisdom."[4]

In human history, the tools we make reflect our embodied state of intelligence. Anthropologists pour over ancient blades, weapons, and agricultural devices to understand past modes of thinking. The extended physical experiences of our forebears shaped the ordinary things we use today. Philosopher John Haugeland explains:

> Think how much "knowledge" is contained in the traditional shape and heft of a hammer, as well as in the muscles and reflexes acquired in learning to use it—though, again, no one need ever have thought of it. Multiply that by our food and hygiene practices, our manner of dress, the layout of buildings, cities, and farms. To be sure, some of this was explicitly figured out, at least once upon a time; but a lot of it wasn't—it just evolved that way (because it worked). Yet a great deal, perhaps even the bulk, of the basic expertise that makes human intelligence what it is, is maintained and brought to bear in these "physical" structures. It is neither stored nor used inside the head of anyone—it's in their bodies and, even more, out there in the world.[5]

Our designs and behaviors arise through and out of interaction with the environment. And the type and extent of this interaction are determined by the body.

At first glance, it seems that our ability to make plans could break this rule of embodiment. Certainly, we can plan a vacation, draw up a new living-room design, or arrive at a political opinion by "the mind alone." Neuroscience, however, reveals a more complex situation. In 1983, Benjamin Libet conducted a series of experiments in which the subject was asked to make the simple decision to move a finger and to record the moment this decision was made. Sensors also recorded the nerve impulse from brain to finger and found that the impulse was on its way roughly half a second before the subjects consciously registered that they were going to move their fingers. Thus, it seems that the choice preceded conscious reasoning. The subconscious mind and the body had things under way before the conscious introspective mind knew about it.[6] Anyone who has driven a car while talking on a cell phone, or while thinking over the day's schedule, knows that the subconscious mind and body can keep things in hand, up to a certain point, while the conscious mind works on other things.

The idea that a computer needs a body has always had its niche in the world of artificial intelligence. Almost every artificially intelligent computer that has appeared in the realm of science fiction has been a robot, often with a more or less human-like body.[7] Rodney Brooks at MIT notes that the basic problems with symbolic AI are rooted in the fact that the problem-solving programs it produces are not situated in the real world. Brooks, and others at a variety of AI labs, have built a series of robots that act within the world on the basis of data acquired through sensors. Brooks began with a series of insects, later moving on to the humanoid robots Cog and Kismet, which acquired some of the rudimentary skills of a baby through interaction with human beings. None of these robots comes close to human-like intelligence, but some seem to have a niche in their environment. Consider the Roomba, a roboticized vacuum cleaner that navigates around a room looking for dirt, avoids furniture and stairs, and plugs itself in when it needs to be recharged.[8] One might argue that Roomba shows as much intelligence as many animals in its ability to navigate in a local environment, avoid hazards, and forage for sustenance.

During World War II, the British mathematician Alan Turing found a way to unscramble German codes, revealing their battle plans to the Allies. Naturally, this brilliant exploit prompted him to think about the nature of intelligence. He concluded that it is a relational process—which is the third kind of approach seen in artificial intelligence today. We guess at someone's intelligence, in other words, by talking to that person. In his 1950 landmark paper "Computing Machinery and Intelligence," Turing presented a test for deciding if a computer is intelligent. In this test, an interrogator slips written questions into a room through a slot and, by the written replies that come back, tries to decide whether the denizen in the room is a human or a computer. If the interrogator fails as often as she succeeds in determining which was the human and which the machine, the machine could be considered as having intelligence.[9] Turing predicted that by the year 2000, "It will be possible to programme computers . . . to make them play the imitation game so well that an average interrogator will not have more than a 70 percent chance of making the right identification after five minutes of questioning."[10] This, like most predictions in AI, was overly optimistic. No computer has yet come close to passing the Turing Test.

Turing was not alone in turning to discourse as a hallmark of intelligence. Discourse is unique among human activities in that it subsumes all other activities within itself, at one remove.[11] Terry Winograd and Fernando Flores assert that cognition is dependent upon both language and relationships. Objects we have no words for do not exist for us in the same way as those we name. We make distinctions through language. Without words to describe difference, distinctions cannot long be held in mind nor shared with others. But discourse is essentially a social activity. The act of speaking to another is not simply the passing of information. Winograd and Flores note, "To be human is to be the kind of being that generates commitments, through speaking and listening. Without our ability to create and accept (or decline) commitments we are acting in a less than fully human way, and we are not fully using language."[12] Understanding arises in listening not to the meaning of individual words, but to the commitments expressed through dialogue. Thus, understanding is both predicated on and produces social ties.

To navigate the world of relationships, one needs what has recently been termed "emotional intelligence." When viewed superficially, emotions seem to obscure rational thought. However, recent research has shown that emotions, far from getting in the way of thought, are actually necessary for cognition. In *Descartes' Error*, Dr. Antonio Damasio notes that patients who have had a brain injury to the parts of the brain that govern the ability to feel emotions also lose the ability to make effective decisions, even decisions as simple as what to have for lunch. Neurophysiologist J. Z. Young notes that "even the simplest act of comparison involves emotional factors."[13] If we have no fears, no desires, we have no reason to value one choice over another. Harvard psychologist Joshua Greene has used brain-imaging techniques to study moral decision making. He notes that our brain automatically generates a negative emotion whenever we contemplate hurting someone. Greene's data suggest that psychopaths cannot think properly because they lack normal emotional responses. "This lack of emotion is what causes the dangerous behavior."[14]

Can a computer have emotions? As computer scientist Marvin Minsky quipped, the question is not whether a machine can have emotions, but whether machines can be intelligent if they do not have emotions.[15] Damasio's and Greene's data show that the answer is no. But how far have we gotten in programming emotions into a machine? Not very far. MIT scientist Rosalind Picard has shown that a computer can be programmed to recognize emotion in either facial expressions or tone of voice. The cute robots in the MIT lab, Cog and Kismet, express a variety of emotions, like fear, amazement, or pleasure.[16] However, while computers can be programmed to recognize or express emotion, actually feeling emotion requires a level of self-consciousness current machines lack.

Turing, Damasio, Winograd, and Flores all view intelligence as based upon some form of social activity. Though they approach it in different ways, each suggests that the idea of an individual intelligence is meaningless; intelligence has meaning only in encounter. Whether a computer could have the capability of entering into true relationship with human beings remains to be seen.

What about the Soul?

In the seventeenth century, the French philosopher René Descartes famously argued that the soul and body are utterly separate. Today, it might seem that the insights of computer-science brain research—with their emphasis on the hardware of the body—would destroy this "Cartesian dualism" once and for all. Oddly enough, this is not the case. The concept of a self separate from the body has been given a recent boost, precisely by computer technology. Many computer enthusiasts today spend large amounts of time interacting in a bodiless world—a "virtual" world that Descartes could hardly imagine. Activities that once took place in real space now take place in cyberspace. Consider: we communicate via chat rooms, text messages, and e-mail; we shop, bank, and do research on the Internet; we amuse ourselves with video games, MP3s, and streamed videos, or as avatars in a "Second Life." We project our minds across vast distances or into fictional realms and have experiences in those places that form us as persons.

This does, of course, have certain advantages. Neal Stephenson, in his novel *Snow Crash,* notes that in cyberspace, "If you're ugly, you can make your avatar beautiful. If you've just gotten out of bed, your avatar can still be wearing beautiful clothes and professionally applied makeup."[17] One can project an image of oneself, and that image is utterly malleable, changed at the flick of a bit.

While the ability to design a body amuses, the greatest seduction of a bodiless existence lies in the fact that our bodies are mortal, subject to sickness, aging, and ultimately death. Computer scientist Ray Kurzweil, in *The Age of Spiritual Machines,* suggests that cyberspace provides a place where we can evade the mortality of the body by downloading our brains into successive generations of computer technology. Kurzweil writes:

> Up until now, our mortality was tied to the longevity of our hardware. When the hardware crashed, that was it. For many of our forebears, the hardware gradually deteriorated before it disintegrated. . . . As we cross the divide to instantiate ourselves into our computational technology, our identity will be based on our evolving mind file. We will be software, not hardware. . . . As software, our mortality will no longer be dependent on the survival of the computing circuitry . . . [as] we periodically port ourselves to the latest, ever more capable "personal" computer. . . . Our immortality will be a matter of being sufficiently careful to make frequent backups.[18]

Kurzweil thinks we might achieve this new platform within the next fifty years. He is not the sole holder of this expectation, though he may be among the more optimistic in his timeline. In *The Physics of Immortality*, physicist Frank Tipler conjectures that the universe will cease to expand and at some point end in a contraction that he calls the "omega point." Tipler sees this omega point as the coalescence of all information, including the information that has made up every person who ever lived. At such a point, the information making up any given individual could be reinstantiated, resulting in a form of resurrection for that person, though Tipler is vague as to how such a reinstantiation might come about.[19]

Both Kurzweil and Tipler hold a worldview that would seem, at first glance, to be at odds with the reductive physicalist position held by many scientists today. Yet their views actually are quite consistent with this worldview. They suggest that the soul is, first of all, nothing more than the collection of memories, experiences, and thoughts that we hold in the neural connections of our brain. In other words, our soul is information. This is seductive for the computer scientist who sees the world in terms of 0s and 1s. Our soul is the information that emerges from the state of consciousness, a quality only held by matter that has evolved or self-organized into a sufficiently complex system.

Biologist Francis Crick expresses this view well: "You, your joys and your sorrows, your memories and your ambitions, your sense of personal identity and free will, are in fact no more than the behavior of a vast assembly of nerve cells and their associated molecules. . . . You're nothing but a pack of neurons."[20] The "you" that Crick speaks of here is not initially disembodied but arises from the workings of the brain. Without such a material basis, "you" cease to exist. But what we identify as "you" is not the brain itself, but the information stored in that brain. In this view, the soul, as information, though dependent on the body initially, later becomes completely separable from the body.

Here we have a first sense of the concept of soul as that part of the self that transcends our mortality. Is doing so on a different platform, such as a computer, consistent with the Christian understanding of the soul's immortality? Not really. The Nicene Creed states that our resurrection is one "of the body," and Paul makes clear that the resurrected body will be a new and different body than our current one (1 Corinthians 15:50). The problem with computer hardware as the platform for this new body is that it is not a part of a new creation but a continuation in this creation. Donald MacKay notes the difference:

> If the concept of creation is to be thought of by any analogy with creation as we ourselves understand it—as, for example, the creation of a space-time in a novel—then a new creation is not just the running on and on of events later in the original novel: it is a different novel. A new creation is a space-time in its own right. Even a human author can both meaningfully and authoritatively say that the new novel has some of the same characters in it as the old. The identity of the individuals in the new novel is for the novelist to determine. So if there is any analogy at all with the concept of a new creation by our divine Creator, what is set before us is the possibility that

> in a new creation the Author brings into being, precisely and identically, some of those whom He came to know in and through His participation in the old creation.[21]

The physicist and theologian Bob Russell once put it to me this way: "Immortality does not just mean more time." After all, we all know—scientists and theologians alike—that the earth is itself temporal and finite, that "heaven and earth will pass away" (Mark 13:31).

This fact of human mortality, and the finiteness of the world, has often been used to explain the source of so many human evils and missteps. In the twentieth century, the Protestant theologian and social critic Reinhold Niebuhr wrestled with the concept of human "sin," especially as it played out in utopian projects of communism and fascism, and even the attempts of democracies to use force and still claim sinless innocence. He traces sin to human finitude. The evils of the world, he goes on, often arise from human attempts to assert powers that belong only to God. We need to accept the finitude of our bodies and minds:

> Man is ignorant and involved in the limitations of a finite mind; but he pretends that he is not limited. He assumes that he can gradually transcend finite limitations until his mind becomes identical with universal mind. All of his intellectual and cultural pursuits, therefore, become infected with the sin of pride.[22]

In this post-Cartesian age, we have come to realize that human beings have a single nature with two inseparable elements, a self-transcending mind and a finite body. Denying the body has led to a worldview that denigrates both the natural environment and women. For if we could live in the bits of a computer, of what use is the natural world? If we can replicate ourselves through back-up copies, who needs babies or even sexual differentiation? Here

I note, however, that, while for them it serves no reasonable purpose, proponents of cybernetic immortality are loath to give up sexual experience itself. Tipler waxes eloquent on the possibility of fulfilling all our sexual desires at his omega point, and Kurzweil is equally enthusiastic about the possibilities of disembodied sexual experience.[23] But these experiences are viewed only in terms of self-gratification, not as true relationship, with all the complexity that that entails.

The relational nature of intelligence suggests that the model of an artificial intelligence that holds a separate identity and acts by itself in the world, as a replacement for human intelligence, is the wrong model. What we truly need are machines that complement what people do, working with human beings to accomplish tasks we cannot do alone. Yet there is one caveat. Human beings are far more flexible than computers. We easily overidentify with and overuse our machines. We all see this in our society's current obsession with quantifiable data. Programmer and commentator Jaron Lanier suggests that, should a computer actually pass the Turing Test, it might not be the case that the computer has become smarter or more human but that our immersion in a computerized world has led humans to become more like machines. Miniature Turing Tests happen whenever we adapt our way of acting or thinking to our software; "we make ourselves stupid in order to make the computer software seem smart."[24] Ethicists Joanna Bryson and Phil Kime have pointed out that our overidentification with computers has led "to an undervaluing of the emotional and aesthetic in our society. Consequences include an unhealthy neglect and denial of emotional experiences."[25] The moral of the story? We can protect the emotional richness of life if we see intelligence as a complex phenomenon, rooted firmly in our bodies as well as minds.

Does an embryo have a soul? When does the soul depart in death? Could a computer have a soul? Perhaps the questions that bedevil us in the twenty-first century do so because they are the wrong questions. The very nature of these questions presupposes the Cartesian dualism of a separate soul, distinguishable from both body and environment. The relational nature of intelligence raises the possibility that thinking of soul in individual terms may be misleading. The mystical sides of many religious traditions have long suggested that we are deeply connected to one another. The nature of this connection is beyond the scope of this chapter. However, our current understanding of intelligence, both human and artificial, tells us that intelligence, consciousness, and yes, probably the soul as well, are meaningless outside of the context of the human being in a web of relationships with other humans and with the environment.

Playing Games in a Virtual World

My hometown, Cold Spring, Minnesota, could be just a few miles from Garrison Keillor's fictional Lake Wobegon. It is the sort of town where neighbors look in on each other and doors are often left unlocked. As in Lake Wobegon, the children are all above average. And in September 2003, one of those children, fifteen-year-old Jason McLaughlin, brought a .22 caliber Colt semiautomatic to school and shot and killed two of his classmates.

As with similar school shootings at Columbine, Paducah, Springfield—other folks' hometowns in Colorado, Kentucky, and Oregon—the media were quick to note that the boys had been avid video game players. This could, however, go without saying. In a survey of 778 students in grades four through twelve conducted in December 2003, the National Institute on Media and the Family (NIMF) found that 87 percent of all students and 96 percent of the boys reported playing video games regularly.[26] A study of students in grades three to five showed that boys play video games an average of 13.4 hours per week, while girls play an average of 5 hours.[27] A second study, of slightly older students in grades eight and nine, showed a continuation of these same average hours of play.[28] Adult players also add to the $10- billion-a-year industry; the average gamer is now twenty-eight years old.

Shortly after the shooting in Cold Spring, I asked one of my classes at a nearby university where I teach computer science about their experience with video games. Pandemonium ensued among the twenty-two young men in the room. Clearly, I had hit on something that interested them far more than computer theory.

Video games are a major factor in the lives of our young people. On one hand, they have been shown to help develop motor skills, certain kinds of logical thinking, and creativity. On the other, researchers say that the sheer excitement of the virtual experience releases chemicals in the brain, potentially teaching the brain to desire more of them. Video games have come a long way since the days of Pac Man or Super Mario. Today they are visually stunning, challengingly complex, and deeply immersive. There are role-playing games, puzzles and strategy games, simulations, and sports games, such as virtual football or skateboarding. However, by far the largest category of games, and the ones my students report that they prefer, are "first-person shooter" games, in which the player faces down other players, monsters, or characters, games that sport names like *Halo, Warcraft, Vice City, Doom, America's Army,* and *Manhunt.* One student noted, "Everything but the sports games requires you to kill."

This killing has become increasingly graphic over the years. Whereas in the 1980s or early 1990s shooting an opponent would have merely resulted in the collapse of that figure on the screen, today's graphics allow for flying gore and body parts, realistic writhing, and screams of pain. "There's blood everywhere," one student succinctly put it. While most games used to come with a "blood off" default setting, today's games are generally "blood on." The faster pace and more powerful weapons in many of these games results not only in more graphic kills but in an increasing number of kills as well. In *Carmageddon,* in which the player runs down pedestrians and crashes into other cars, one researcher estimated that the player would have killed nearly 33,000 people if he had completed all levels of the game.[29]

As a response to this increase in violence, the Entertainment Software Rating Board has developed a rating system designed to keep the most violent games out of the hands of young children. Unfortunately, this system is little understood by parents and often unenforced by vendors. The highest rating, AO (adults only), is designated for games that "include graphic depictions of sex and/or violence." However, since most major retailers will not sell AO games, this rating is almost never used. Games that clearly include both graphic sex and violence, such as *Grand Theft Auto* or *Manhunt*, are rated M (not appropriate for persons under 17). In the NIMF survey quoted above, 87 percent of boys in grades four through twelve reported playing M-rated games, with 78 percent ranking M-rated games among their top five favorites. Politicians around the country have introduced a variety of bills seeking to regulate the sale of violent video games to children, but so far no such laws have passed both the legislature and the courts.

Only a Game?

But it is only a game, right? Yes and no. Killing a fictional character does not cross many moral boundaries. In multiplayer games, the character may represent another player, but you have not killed that player in the real world. The real issue is not what the violence may do to another, but what it does to the player himself. A growing body of studies of a variety of types—laboratory studies, field studies, longitudinal studies, and cross-sectional correlation studies—shows growing evidence that playing violent video games results in increased aggressive behavior. A Japanese study of fifth and sixth graders showed a clear correlation between the amount of time spent on video game playing early in the year and later physical aggression.[30] Three recent American studies found a similar link between violent game playing and aggressive thoughts and behavior, even after controlling for innate temperament and exposure to violence in other sources, such as movies and television.[31] Dr. Craig Anderson of Iowa State University, surveying these and prior studies, notes clear evidence that playing violent video games results in increased aggressive thoughts, feelings, and actions and decreases in helping behavior.[32]

The results of these studies should be no surprise. After all, these games reward the player for mastering violence. Moreover, they teach that violence is an appropriate response to threat. That is the reason these games are used by the military as recruiting and training devices. *America's Army,* a first-person shooter game, is both distributed on CD by Army recruiters and downloadable from the Army's web site. The game's basic training level has been completed by more than 4.5 million players. The game *Doom* has also been used by the Marine Corps as a training device. Lt. David Grossman, retired professor of psychology at West Point, notes that these games provide a script for the rehearsal of the act of killing: "It is their job to condition and enable people to kill. . . . [These games] teach a person how to look another person in the eye and snuff their life out."[33]

Perhaps the more disturbing thing about the relationship between the military and video games, however, is not that games like *America's Army* imitate war but that war seems to be increasingly imitating video games. The Bush administration's policy of preemptive attack, underlying the war in Iraq, is a given in the video-game world, a world in which the "bad guys" must be killed before they kill you. Video games teach quick reaction, not reasoned response. Retired Army general Wesley Clark, in an article for the *New York Review of Books,* noted a second problem with the war in Iraq that seems oddly video-game related—the initial, almost exclusive focus on hitting targets rather than planning for the difficult task of building a country. Clark describes the Defense Department's vision for Operation Iraqi Freedom as the detection and destruction of enemy forces with minimum risk to one's own forces, emphasizing dominance through precision strikes. In the words of one senior officer, "Imagine a box of enemy territory 200 kilometers wide and 200 kilometers deep; we should be able to detect every enemy target there, and to strike and kill any target we want."[34] Sounds a lot like a video game.

The same thinking can strike closer to home. Chris Magnus, police chief of Richmond, California, attributes the recent rise in violent crime in the United States not merely to gangs, to easy access to guns and drugs, and to poverty but also, in part, to video game playing: "There's a mentality among some people that they're living some really violent video game. . . . We seem to be dealing with people who have zero conflict resolution skills."[35] Video games do teach other skills and improve human performance in some areas. A 2003 study at the University of Rochester, New York, showed that proficient gamers markedly improved their spatial skills and their ability to pay attention to changes in the visual environment.[36] Reaction time is faster in video game players. The games also promote pattern matching and systematic thinking. Video games have been used to help recover mobility in stroke victims and to overcome fears in the phobic. The Federation of American Scientists has endorsed video games as a means of teaching "higher-order thinking skills, such as strategic thinking, interpretive analysis, problem solving, plan formulation and execution, and adaptation to rapid change."[37]

One reason cyberspace is an excellent venue for learning skills is that it offers the opportunity for endless repetition of a task. However, an ethical issue arises when one considers that the tasks repeated by many video game players are violent ones. In a study of a random sample of video games rated T (for Teen) by the Entertainment Software Rating Board, 98 percent included violence, with an average of 122 deaths per hour of game play. Sixty-nine percent either rewarded a player for killing or required a player to kill.[38]

Such first-person shooter games improve marksmanship, which is another reason they are used by the U.S. military for training. But most are used by teens. Attorney Michael Breen, representing families of three students killed in a school shooting in Paducah, Kentucky, noted their efficacy:

> [The shooter] clipped off nine shots in about a 10-second period. Eight of those shots were hits. Three were head and neck shots and were kills. That is way beyond the military standard for expert marksmanship. This was a kid who had never fired a pistol in his life, but because of his obsession with computer games had turned himself into an expert marksman.[39]

Once again, is it really only a game? Evidently not to our brains. Dr. Klaus Mathiak at the University of Aachen used MRI technology to study the brains of thirteen men who played violent video games for an average of two hours a day. He found that, during the fights in the video game, the emotional centers of the brain associated not only with aggression but also the amygdala and parts of the anterior cingulate cortex, became more active. This pattern is associated with aggression and with the suppression of positive social emotions such as empathy. Mathiak speculates that playing violent video games would strengthen these patterns in the brain over time: "Contrary to what the industry says, it appears to be more than just a game."[40] As educational psychologist Jane Healy puts it, "Habits of the mind become structures of the brain."[41] Immersion in cyberspace sets up a cybernetic loop between the human and the machine, a loop that allows each to be changed by the other. The player plays the game, and the game plays the player.[42] The rewards system in video games also triggers the release of dopamine. Dopamine triggers both learning and satisfaction. This, of course, may underlie the addictive potentialities of these games, which seem to affect roughly 10 percent of players.[43]

Living in Cyberspace

Cyberspace also gives an illusion of human enhancement. Gamers report feeling empowered, freed from the structures of normal life. In particular, in the virtual world, one is freed from the limitations of the human body; one can move in three dimensions, into and through objects, and can appear and disappear. Software engineer Michael Benedikt envisions cyberspace as a place where "we would enjoy triumphs without risks and eat of the tree and not be punished, consort daily with angels, enter heaven now and not die. . . . [It is] the Heavenly City, the New Jerusalem of the Book of Revelation. Like a bejeweled, weightless palace it comes out of heaven itself . . . a place where we might re-enter God's graces . . . laid out like a beautiful equation."[44] Game players experience this freedom in their abilities to leap, fly, and float through space, walk through objects, and fire guns without reloading.

The ultimate human enhancement is not to die. Virtual immortality is simulated through the option of playing in "God mode," in which the player becomes invincible, is given unlimited weapons, special powers, or unlimited lives.[45] In the words of one player, "I really like games—especially shooting games—that have some kind of invincibility option or god-mode, and you get to just run around and cause total [bleeping] havoc. Explosions and limbs flying everywhere. Ahhhh . . . Sometimes that's just what you need."[46] Of course, God mode is over once the game is over.

No one can live in cyberspace forever. And the more one lives in cyberspace, the less one lives in the real world. Hours spent in front of a screen are hours not spent messing around outdoors. Carl Pope, director of the Sierra Club, describes what is lost: "In losing our contact with the natural world we are losing something precious. In a way, we are losing part of what it means to be human. We evolved in nature, dependent on its rhythms, inextricably connected to other living things. . . . American children are losing that connection."[47]

From a religious perspective, there are obvious reasons to be concerned about the increased prevalence of violent and graphic video games. Exposure to simulated violence and death can desensitize, lowering inhibitions and making it easier to commit violence in the real world. Violence is romanticized and equated with personal power and achievement. Many games include a "back story" that explains the characters and their motivations. Revenge is a common feature in these stories, fostering the notion that violence as payback is seemingly justifiable. Nick Yee, research scientist at Palo Alto Research Center, notes, "It's hard to have an in-game and out-game moral compass. I think it's the same thing, and when you play the game, your moral compass gets influenced and impacted by your decisions."[48] Though one does not kill real people, one gets used to the concept of killing. The field of virtue ethics warns us that one's character is formed by one's habits.

First-person shooter games are designed to present the world in adversarial terms and to inure the player to violence. But what about games that are not violent? There are two kinds worth considering: simulation games and role-playing games.

In simulation games, one builds or controls a virtual environment. Examples include *SimCity,* in which cybercharacters go about their day-to-day lives; *Tropico,* in which you can run a banana republic; or *RollerCoaster Tycoon,* in which you build and run your own theme park. One can even simulate an ant colony with *SimAnt.* In other constructive games, one can run a nongovernmental organization or even a government. These games are all about making choices, and call on the player to assert herself, to exercise her will in manipulating the virtual world. Though not as obviously as in the first-person shooters, the player is still the actor and holds her power alone, not in a web of relationship with others.[49]

As we saw in the first chapter, the theologian Karl Barth elaborated the Christian idea that when humans exist in the image of a trinitarian God, full humanity is achieved only in relationship with others. This vision of humanity places encounter and cooperation at the center. Is this vision a part of the video game world? Role-playing games, often designed in an adventure/quest format, allow, and often require, interaction and cooperation between players. *EverQuest* is one of the most popular massively multiplayer role-playing games. Players create their own characters, go on quests, solve puzzles, and kill evil creatures. Other players can be either teammates or opponents, and some of the quests can be solved only in groups. However, such cooperation is rare in the video game world. Most games demand that the player act alone. Even in *EverQuest*, the players must often make decisions too quickly for them to be in any real sense collaborative. In the end, the cyber-world of video games can be a very lonely place. As David Grossman notes, "Most children who are traumatized and brutalized through their exposure to violent media do not become violent, but they do become depressed and fearful."[50]

The world of video games is not only a lonely place, but a rather simple place as well, lacking the complexity of the real world. Video-game characters tend to be black or white. They rarely develop or exhibit any of the complexities of real human beings. And, while actions in the game world do have consequences, these too tend to be simple and static. If you walk around a certain corner, you will get shot. Every time. If you fail to maintain the buildings in your simulated city, some will fall down. Every time. The consequences are inside the game; they have little relevance to the complexity of the real world. Thus, while a video game player gains experience, it is experience bounded by the limited and, with enough playing, predictable world of the game.

Many of my students spoke of playing video games as an occasional release, a way to get together with the guys and hang out, a chance to get an adrenaline rush in a safe way. In this sense, they provide entertainment and enjoyment. As an occasional pastime, video games seem harmless enough. But when the average American child spends nine hours a week playing these games, we need to ask what sort of worldview the games are furthering.

We are a narrative people, passing on our values, our abilities, and our faith through the stories we tell one another. What do these games tell our children about what it means to be fully human, about decision making, about social roles, about living in the real world? Eugene Provenzo, professor of education at the University of Miami, considered these questions in his testimony before a Senate committee hearing on interactive violence and children as follows: "[These games] are the cultural equivalent of genetic engineering, except that in this experiment, even more than the other one, we will be the potential new hybrids, the two-pound mice. It is very possible, that the people killed in the last few years as the result of 'school shootings' may in fact be the first victims/results of this experiment."[51] In my hometown, Jason McLaughlin acted out a vision of the human being as lone actor in a hostile world. His response to persistent teasing was to act out the role of first-person shooter. What influence video games had on his view of the world we may never know. But tragedies such as this call us to be thoughtful about how we let our children fill their hours and their minds.

Relationships in Cyberspace

Shy and awkward, thirteen-year-old Megan Meier thought she had a new friend, a boy called Josh Evans, whom she met on MySpace. But there was no Josh Evans; his MySpace profile was a hoax, initiated by the mother of a girl who had once been a friend of Megan's and perpetuated through postings by her and others, postings in which "Josh" moved from flirtation and fun to insult and criticism. On October 16, 2006, after receiving a message from "Josh" that said, "I don't like the way you treat your friends and I don't know if I want to be friends with you. . . . The world would be a better place without you," Megan fled to her bedroom and hanged herself. She died the next day.[52]

If video games have become the electronic world of young men, on-line social networking has come to play a large role in the social world of young women. MySpace is only one of several sites where a user can create a profile, post comments and photos, designate other users as "friends," and share audio and video files. It is estimated that MySpace has over 100 million members, most in their teens or twenties. A similar network, Facebook, claims more than 60 million users in generally the same age group.[53] A recent survey conducted by the Pew Internet and American Life Project found that 64 percent of American high-school students, both boys and girls, have posted content on some on-line social networking site.[54]

Social networking sites are only one on-line venue for communication. Others include e-mail, text messaging, instant messaging, chat rooms, web pages and web cams, and multiuser domains. Virtual reality programs immerse us in new realms. These new technologies have been hailed as a revolution in both our ability to communicate and our ability to experience. If so, they are significant not only in our day-to-day experiences but to religion, for our ability to relate, whether with one another or with God, lies at the heart of our religious experience. Do these new media of communication really represent a breakthrough in our ability to relate to one another? Do they offer a new venue for religious experience? Or are these illusions, hyped by Silicon Valley entrepreneurs and the writers at *Wired*?

Virtual Relationships with Others

Language is foundational to our ability to relate to one another. As communications scholar Quentin Schulze points out, "We are creatures of the spoken word."[55] The development of oral communication marked the most important shift in human evolution. It allowed us to develop communal memory and to pass what had been learned on to future generations, making human evolution as much Lamarckian (passing on learned traits) as Darwinian (passing on genes). Each of the technologies of writing, printing, telegraph, telephone, radio, television, and now computer has changed the nature of oral communication, most often by changing its reach, but has not improved upon it. New technologies add but also take away from our ability to converse with another. For example, the telephone allows communication at a distance but takes away the input of facial expression and body language. Television allows both visual and aural communication with an incredibly large number of persons, but this communication is only one way. Most forms of technologically mediated communication are in one dimension or another more static than speaking face to face. Does this matter? Marshall McLuhan was remarkably prescient when he wrote in 1964,

> The medium is the message. This is merely to say that the personal and social consequences of any medium—that is, any extension of ourselves—result from the new scale that is introduced into our affairs by each extension of ourselves, or by any new technology.[56]

McLuhan notes that every extension of our abilities, particularly a technological extension, either amputates or modifies some other aspect of that ability. In the process of adopting a new technology, we tend to focus on the extension that technology gives us while ignoring or minimizing those aspects that are compromised.

In the case of on-line communication, what is extended is both reach and scale. E-mail lets one communicate within minutes with friends and colleagues around the globe. Web pages, such as those on Facebook, allow words and images to be transmitted almost instantaneously to anyone with a computer. The character I am speaking to or texting in Second Life might be the avatar of someone in China or Finland. Physical distance is no barrier in cyberspace. Similarly, on-line communication greatly increases the number of people one can reach. I can send an e-mail to all of my friends and colleagues with the click of a single button. A writer who might have had an audience of the few thousand people who read her local newspaper can now post a blog with the potential of millions of readers around the globe. A high-school student can "text" his friend on his cell phone in the middle of biology class. Clearly, we have never had so many options for verbal communication.

One thing that is often compromised, however, is authenticity. On the one hand, sites such as MySpace or Facebook seem to encourage utter transparency. It is rather amazing what sorts of personal information many users post, from intimate musings to pictures of drinking binges or romantic encounters. Researchers Alessandro Acquisti and Ralph Gross, who surveyed Facebook users at Carnegie Mellon, comment: "One cannot help but marvel at the amount, detail, and nature of the personal information some users provide, and ponder how informed this information sharing can be."[57] For young users, such transparency can be dangerous. Acquisti and Gross found that 40 percent of the users they surveyed posted their class schedule, and about one in four posted his or her address. Other users post their cell phone numbers, e-mail address, and even their sexual orientation. One-third of teens using social networks reported that they had been contacted by a total stranger and a significant number felt threatened by this contact.[58] Recently, MySpace removed the accounts of over 90,000 known sex offenders from their site.

For older users, misjudgment regarding transparency carries other kinds of risks. A survey of employers conducted at the University of Dayton found that 40 percent said they would consider using a Facebook or MySpace profile as one source of information in their hiring process, and several reported rescinding an offer of employment based on what they saw on the Web.[59] College deans and advisors have disciplined students for underage drinking or other infractions that they found advertised on students' pages.[60] Even if a student himself would never think of posting such a picture on his own page, anyone with a cell phone at a party might snap a picture and post it, often without thinking of the consequences.

On the other hand, the tendency toward self-exposure and over-transparency on these sites also encourages dissemblance. As the sad story of Megan Meier points out, it is easy to present oneself as something or someone one is not on the Web. Cyberculture even celebrates the idea that in cyberspace one can explore aspects of one's personality one would not present publicly in the real world. Thus, a man might design a female avatar for himself in a venue such as Second Life, or a woman might check out what it feels like to enter into a lesbian relationship. Age is one of the most frequently lied-about characteristics, as the young attempt an older, more experienced identity, while others relive their youth. At thirteen, Megan Meier was too young to legally have a MySpace page; she presented herself as sixteen in her screen name. In a natural desire to be noticed and popular, young people look for ways to set themselves apart. Accomplishments and personal quirks are all too easy to exaggerate, and exaggeration can quickly turn into duplicity. Gossip, which might be checked if passed in the real world, gains a life of its own. Rushworth Kidder, founder of the Institute for Global Ethics, notes,

> What's lost in the shift is a kind of reciprocal correction mechanism. What's missing is the in-person pushback from friends and adults that operates through thousands of visible and linguistic hints, telling young people what works and what doesn't. Calibrating their identities along a scale of courage that runs from timidity to bravura, young people rightly ask themselves, "Am I too brash or too fearful?" Face-to-face interaction sends back highly nuanced and helpful answers.[61]

Without this feedback, one ends up with political scientist Francis Fukuyama's "culture of unbridled individualism, in which the breaking of rules becomes, in a sense, the only remaining rule."[62] Although the goal is individuality, a quick look at a selection of MySpace or Facebook pages shows a sad sort of sameness: pictures in goofy poses, sound bites in text message style, shortened spellings, quasi-sexy icons and inserts.

A second loss lies in the obvious conclusion that time spent on-line means time not spent doing other things. Two thirds of Facebook members log on every day, and many users spend hours on the site, browsing through other's pages or perfecting one's own. A young woman who spends her evening on Facebook is not spending that time interacting directly with family or friends. Rather than hanging out in the living room or at the mall, she is alone in her room. In a two-year study of high-school students, researchers found that those who increased their Internet use not only decreased time with family and friends but also showed a greater tendency toward loneliness and depression.[63]

Clearly enough, "friends" on a social network do not play the same role as friends in the physical world. On-line communication is often truncated to short posts, text messages, or e-mail. It cannot occur as quickly as in a direct conversation. It also tends to be more superficial. Without the clues given by tone of voice, body language, and the clarifications that are easy to make quickly in a face-to-face conversation, we are reluctant to move to the same level of intimacy and self-revelation. Not that intimacy cannot be initiated or sustained on the Web; the sheer number of Internet dating services indicates that it can. But many use social-network sites as a way to avoid the hard work that intimate relationships demand. A recent Harvard graduate noted of Facebook, "It's a way of maintaining a friendship without having to make any effort whatsoever."[64]

While the computer does offer the opportunity to widen one's circle of acquaintance, in another way it often has a limiting effect. Real-world communities are made up of persons with differing views and backgrounds. Physical proximity forces us to interact with people we might not choose as friends. The Internet facilitates a more general movement in our society away from neighborhoods, communities defined by physical proximity, toward what sociologist Robert Bellah calls "lifestyle enclaves."[65] On-line, most people gravitate to groups that already share their interests or worldview. This has led to a polarization in on-line communities, particularly in the political arena. Liberal blogs are read by other liberals, while conservatives gather with their own. Quentin Schulze, a communications scholar, notes that the Internet fosters individualism at the expense of community: "Cyberspace empowers us to become more efficient libertines, scurrying along the digital trails in search of anything that we desire, dropping our two cents worth like mouse pellets along the way. . . . For many people, the real lure of cyberspace is personal expression, not mutuality."[66]

For many critics, this personal expression is a part of a larger increasingly narcissistic culture. Sociologist Duncan Watts notes, "If I had to guess why sites like Facebook are so popular, I would say it doesn't have anything to do with networking at all. It's voyeurism and exhibitionism. People like to express themselves, and they are curious about other people."[67]

Psychologist Jean Twenge has compared levels of self-absorption among college students over the past twenty years. Today's students are much more narcissistic than previous generations, and Twenge lays partial blame for this on social networking: "By its very name, MySpace encourages attention-seeking, as does YouTube, whose slogan is 'Broadcast Yourself.'"[68]

Broadcasting oneself does not lead toward deep friendship, which requires a certain amount of privacy. Normally, we call someone a friend precisely because we share something with that person that we do not share with others. Yet, for too many, "friends" in a social network is a numbers game indulged in to promote one's social status. According to Christine Rosen, who writes on technology and culture, "Friendship on these sites focuses a great deal on collecting, managing, and ranking the people you know. Everything about MySpace, for example, is designed to encourage users to gather as many friends as possible, as though friendship were philately."[69] Many MySpace and Facebook users list hundreds of friends. Sites also encourage users to rank their friends, adding another venue for the cliquishness and popularity contests all too common in adolescence.

Careful and loving relationships with others lie at the heart of all religious traditions. On-line communication can be one way of maintaining and fostering such relationships. However, young people in particular need to be taught to use the medium thoughtfully and not to overuse it. Truly intimate relationships are still best fostered through face-to-face contact. The theologian Karl Barth sees face-to-face contact as the first ingredient among four in an authentic relationship—to look the other person in the eye and thus acknowledge the other as distinct from one's self. Relationship has a spirit of mutuality, not status seeking. This spirit of self-disclosure to the other is not encouraged on the Web. Second, Barth says, we must speak to, and hear, one another. On-line communication does this—but only sort of. It lacks the actual sound of the other's voice, making it hard to truly "hear" all that comes with human words. The third characteristic of relationship is helping others and doing this gladly.[70] This often requires proximity. It also makes people vulnerable to risk. Rosen quotes from a young woman on this matter of distance: "I consistently trade actual human contact for the more reliable high of smiles on MySpace, winks on Match.com, and pokes on Facebook."[71]

Virtual Relationship with God

In Moscow, the Tretyakov Gallery holds one of the most famous icons in the world, a mystical face of Christ painted in the Eastern Orthodox tradition by Andrei Rublyov. On a visit, you can see people standing or seated, staring at the image for great lengths of time. It may be a stretch to liken contemplation of a Rublyov to someone staring at a flickering computer screen at home, yet it is a comparison that will help us consider whether virtual reality allows for a relationship to God. As we shall see, image is the real power of cyberspace, and image has long been a vehicle for communing with the divine.

The spiritual or religious world is not a total stranger to the Internet. There are web sites for a wide variety of religious faiths and denominations where one can access religious texts, share experiences and prayer requests, build new spiritual friendships or communities, engage in ecumenical dialogue, and even participate in some religious rituals.[72] Web cams let one make a virtual pilgrimage to Mecca, the Wailing Wall, or Chartres Cathedral. Each of these uses draws on the strength of the Internet as a medium that overcomes distance and time. However, as with human–human relationship, what computer technology gives to religion in terms of speed and broader access it takes away through lack of physical presence.

We can certainly find religious texts on the Web. There are also sites where people of different faiths gather for discussion, to ask and answer questions. But these things can often be better accomplished through books or in face-to-face gatherings of religious communities. The greatest strength of the Internet as a portal for spirituality may lie not in the ability of the computer to store and transmit information but in its ability to communicate without words, through images. One of the strongest nontextual uses of the computer is found in the realm of the computer-generated, multidimensional world of virtual reality.[73] Virtual reality involves a projection of the human consciousness into the world of the computer, a world governed by information and mediated through screen, headset and fiber optic glove, audio speakers, or any of a host of other input/output devices. While involvement in such a virtual reality may be purely passive, much like watching a three-dimensional movie, cyberspace is at its best when encountered in an interactive mode. In *Cybergrace*, Jennifer Cobb predicts this interactivity has the potential to engage us in spiritually transformative experiences.[74] According to David Porush, "Humans inevitably feel that a certain architecture is needed to summon the transcendent into this world. . . . For postmodernism, cyberspace is it."[75]

This optimism is, however, open to debate. When we surround ourselves with our own artifacts, we all too often cut ourselves off from nature or from each other. Fr. John Chryssavgis describes this propensity: "Our generation . . . is characterized by a sense of autism toward the natural cosmos, by a lack of awareness of our communication with the beyond. We appear to be inexorably locked within the confines of our individual concerns," and our preoccupation with the artifacts we have constructed contributes to this confinement.[76] Artifacts do not make good windows into the world of the sacred. Rather than turning us outward toward God, they wall us in, locking us in a world of our own making.

The icon, however, is in a different class. Icons have long been considered windows onto the transcendent. They have played a historic role in the spiritual practice of the Orthodox, and, increasingly, other Christians as well. For the Orthodox, they are objects of both veneration and contemplation, used in liturgical and private settings. Icons have a sacramental quality. Though clearly rooted as material objects in the material world, icons open for the viewer a portal into the kingdom of God. Consider this account from an expatriot American who regularly visits the Tretyakov Gallery, heading straight for the room with three large Rublyov paintings, tattered as they may be:

> The face of Christ draws me more powerfully than any other and I think it is because I feel that I am in the presence of a true embodiment of spirit in flesh. There is little left of the painted image, but the expression of the eyes and the face generally are today surely just as Rublyov rendered them. And the effect is just as you say—I feel soothed, welcomed, transported away from the mundane. When sitting before Rublyov's face of Christ, the last thing I think about is myself. I am drawn out and away, into something that I cannot really describe, but it is a wonderful place to be, for I feel that there, I am not alone. . . . I see the image clearly in my mind's eye as I write, all the suffering of human life, all the peace and quiet bliss of the eternal spirit.[77]

In these words, we can find three experiences facilitated by the icon—immersion, encounter, and transformation. The writer notes that he is drawn out of himself and into the world of the icon. Second, he feels he is truly in the presence of Christ, "not alone." Finally, the experience of sitting before the icon effects a change in him; he notes an increased awareness of both the suffering in life and the peace of the spirit.

According to this testimony, the icon takes the beholder into a separate world, a world in which our usual perception of the dimensions of time and space is altered. Immersed in this separate space, the viewer can set aside the busyness of the mind and focus attention on the persons and scenes. Icons are not objects to be enjoyed merely for their beauty nor to be dissected and understood rationally. They call for a viewer's focused attention, and in this focused attention (a form of attention held sacred in almost all of the world's religious traditions), they facilitate movement into a new state of mind. Entering a virtual world is also an immersive experience. Bodily existence alters or disappears, and the mind can roam at will. Nicole Stenger describes this experience: "Cyberspace grafts a new nature of reality on our everyday life. It opens up an infinity of space in an eternity of light. . . . On the other side of our data gloves, we become creatures of colored light in motion, pulsing with golden particles. . . ."[78]

Like the icon, the world of virtual reality is both like and unlike normal Euclidean space; the usual laws of space and time may or may not apply. In virtual reality, one can move in three dimensions, into and through objects, which can be made to appear or disappear. The programmer determines the laws of the virtual world, and those laws need not conform to the laws of the real world. So far, so good. There is, however, one difference between the virtual world and the icon. Jaron Lanier notes, "The biggest problem in computer culture is that on the whole it has not been beautiful."[79] Cyberspace has more often been a place where dreams of violence or eroticism are acted out. It is a place where, all too often, "ethical awareness shrinks and rudeness enters."[80] This is not to say that a virtual world could not be beautiful, only that most, so far, have not been deliberately designed to be so.[81]

Icons always depict persons. As Orthodox priest Metropolitan Gennadios Limouris explains, "Icons, in general, are visible portraits of a deified humanity, of men and women who have recovered the capacity to show forth the divine which had been obscured by sin."[82] The icon is venerated not for itself but for its power to lead to an encounter, first with Christ or the saints depicted, then, through them, to an encounter with the invisible and eternal God. What do we do in this encounter? There is a story in the *Sayings of the Desert Fathers* that relates how the devil, on coming across a monk sitting outside his cell, asks the monk what he is doing. The monk replies, "Nothing. I am simply keeping this place."[83] Icons ask us to cease our constant preoccupation with our own agency, to "be still, and know that I am God."

This is a bit different in the world of virtual reality, where encounters are generally an opportunity to exercise personal agency. Janet Murray writes, "The more realized the immersive environment, the more active we want to be within it. When the things we do bring tangible results, we experience the second characteristic delight of electronic environments—the sense of agency."[84] Rather than calling the user to "be still," cyberspace entices the user to see what might lie around the next virtual corner. Focused attention is difficult to maintain in a world that invites continual movement and change. Computers offer us countless ways of escaping our present body and present circumstances. In cyberspace, one can be what one is not—a man can pose as a woman, a human as an animal. We can reduce ourselves to points of light, and we can pretend to godlike powers or knowledge. But eventually one must remove the helmet and the data glove and turn off the computer screen; then one is precisely what one has always been, man or woman, human, and mortal.

If a true spiritual transformation is what people seek, can it be produced by virtual reality? The characteristics that distinguish virtual reality from regular computer applications are not entirely dissimilar from those that distinguish icons from regular paintings. Both have strong immersive power, through the use of an altered yet familiar reality. Both help us let go of everyday preoccupations and fully enter into a new experience. However, just because an experience is new, that does not guarantee the Divine Presence. For virtual reality to be an aid to religious contemplation, the world it draws one into must be one of beauty, deliberately created with the ideals of some spiritual framework in mind. It must also become a world of encounter. The common model of the user as the lone inhabitant of the virtual world is not a good model for religious contemplation, for it turns the contemplative inward on his or her self and his or her capacities.

Computers are an obviously useful tool for the human community. The benefits they give us in data storage, calculation, and the freeing of the imagination are great. But when we come to the second and third questions the Amish ask about technology, we find some room for discussion. While computers extend the community we can be in touch with, even beyond the human in the case of artificial intelligence, they also change the nature of that community. For artificial intelligence, video games, social networks, and virtual reality to be unequivocal goods in our society, they need to allow for true human relationships, the kind that lift people out of narcissistic preoccupations. That is a tall order. But not an impossible one.

CHAPTER 4

The New Alchemy

A FEW CENTURIES AGO, the German theologian Martin Luther raised a technological topic during his famous "table talks," and later in London, the scientist Isaac Newton pondered the same subject. They called it "alchemy," the power to change matter itself. "The science of alchemy I like very well," Luther said, probably over a pint of Wittenberg ale. "I like it not only for the profits it brings in melting metals, in decocting, preparing, and extracting and distilling herbs, roots; I like it also for the sake of the allegory and secret signification."[1] The dream, of course, was to change base metals into precious gold. In many ways, that dream has not died. On the border of France and Switzerland, the largest machine in the world —the seventeen-mile circular Large Hadron Collider—despite current difficulties, is designed to crash two particle beams into each other at the highest rates of energy ever achieved. The goal is to "decoct" matter itself into its smallest parts and, in the process, to discover the "secrets" of the universe.

In our long history of inventing technologies of matter, the goal has always been to control the natural world around us. As the philosopher Jacques Ellul reminded us earlier in these pages, technology can isolate us from nature; today we have a wall of televisions, computers, BlackBerrys, cell phones, cars, and other machines between us and the great outdoors. Nevertheless, technology is essentially about control of nature. It is about taking hold of matter and wrenching from it what we want, which is exactly what the "new alchemy" is about. This chapter—whose theme is technologies of matter—looks at how we are applying the dreams of alchemy to nanotechnology, genetically modified foods, and our use of fossil fuels and nuclear energy.

Our modern approach to the technologies of matter has one significant difference from that of our forebears, according to historians of science and culture. In the past, the natural philosophers used to see life as embedded within material objects, a view called "animism" or "vitalism." Religious traditions such as Christianity have espoused the idea of stewardship, of care for these natural things as if they were living partners in the Creation. However, the modern mentality sees objects as nothing more than chunks of matter to be used by human ingenuity. In fact, the biblical idea of "dominion" can easily be read in this utilitarian way. In 1967, historian Lynn White stirred a great controversy by pointing this out in a widely read article in *Science* magazine:

> In Antiquity every tree, every spring, every stream, every hill had its own genius loci, its guardian spirit. These spirits were accessible to men, but were very unlike men; centaurs, fauns, and mermaids show their ambivalence. Before one cut a tree, mined a mountain, or dammed a brook, it was important to placate the spirit in charge of that particular situation, and to keep it placated. By destroying pagan animism, Christianity made it possible to exploit nature in a mood of indifference to the feelings of natural objects.[2]

Few today worry about the feelings of natural objects. The ecological crisis of today, however, may be changing our indifference to the natural world. Our technologies have become more powerful, and we are seeing more clearly their side effects, both on nature and society. This is a good time to recall the questions our religious traditions have provided for assessing our technologies: First, does the technology provide tangible benefits to the community or individuals within that community? Second, does the technology change the relationship of the individual to the community? Third, does the technology change the nature of the community? When it comes to technologies of matter—which seek to change nature itself—the community is large. We are challenged to consider not only how revolutions in technology affect the poor and marginal in the world, but even how they affect other species and the very planet itself.

Nanotechnology

As we know, the scientists and philosophers of Luther's and Newton's eras never did transmute base metal into gold, nor has anyone since. But we still continue the quest for what they called the "philosopher's stone," which stands for a material, process, or elixir with extraordinary powers to alter nature. To its credit, alchemy did produce a wide array of useful chemicals and pharmaceuticals. The new field of nanotechnology has likewise produced a variety of new and improved products and promises more. But the similarity between alchemy and nanotechnology does not stop there. Nanotechnology pursues its own philosopher's stone. Its enthusiasts promise us the power to change one form of matter into another, bringing wealth, restoration of the body—and even eternal life.

Materials Technology: Tennis Balls and Sunscreens

Nanotechnology is the design and production of devices or materials on the scale of 1 to 100 nanometers, where a nanometer is one-billionth of a meter. To place this scale in perspective, human hair ranges in width from 25,000 to 50,000 nanometers, a DNA molecule is 2.5 nanometers, and most proteins, 1 to 15 nm. At the nanometer scale, one is essentially working at the molecular or atomic level. Techniques in nanotechnology could ultimately allow us to design anything at that level, attaching whatever molecules we like so long as their attachment is consistent with the laws of physics. This would allow the production of materials that are not found in nature, materials that are lighter, stronger, and more precisely tailored than we can currently produce. It could also give us machines small enough to navigate the human bloodstream.

Nanotechnology represents the newest frontier of both materials science and medicine, on which both businesses and governments are anxious to take the lead. It is less a field in itself than a subfield in each of the fields of physics, chemistry, engineering, computer science, biology, and genetics. The wide diversity of applications in nanotechnology allowed the field to sweep the Nobel Prize competition in 2007, in which the prizes in medicine, physics, and chemistry all involved research at the molecular or nanoscale level. As the newest frontier in both science and engineering, nanotechnology has garnered extreme private and public interest and support. The National Nanotechnology Initiative in the United States has gone from a budget of approximately $500 million at its founding in 2001 to approximately $1.3 billion of governmental funding in 2006; internationally, nanotechnology research dollars totaled $9.6 billion in 2006. The total value of all products incorporating nanotechnology was estimated to be $50 billion in 2006 and is projected to reach $2.6 trillion by 2014.[3]

Ideally, nanotechnology would let us assemble any substance we wish, snapping together the basic molecules like so many Lego blocks. This would require tools that allow us to manipulate individual atoms and molecules. The first such tool was the scanning tunneling microscope, for which Gerd Binig and Heinrich Rohrer received the Nobel Prize in physics in 1981. This microscope, and subsequent, similar tools, allows researchers to "see" at the molecular level, to manipulate molecules, albeit so far only in a two-dimensional way, and to see how nanoscale products interact with cells and other substances. To manipulate molecules on a three-dimensional level, the optimal such tool would be a molecular robot, one that is itself molecular in scale, programmed to manipulate objects at its own level. Physicist Richard Feynman introduced the idea of nanobots in 1959: "I want to build a billion tiny factories, models of each other, which are manufacturing simultaneously. . . . The principles of physics, as far as I can see, do not speak against the possibility of maneuvering things atom by atom. It is not an attempt to violate any laws; it is something, in principle, that can be done; but in practice, it has not been done because we are too big."[4]

The machines envisioned by Feynman do not yet exist. Current nanomaterials have been produced in a bottom-up fashion by relying on natural chemical processes to do the assembly for us. These nanosubstances spontaneously self-assemble when triggered by some change in their environment, such as acidity, temperature, or the application of an electric charge.[5] One of the most celebrated nanomaterials, carbon nanospheres (or nanotubes also known as buckminsterfullerenes or buckyballs) is constructed through molecular self-assembly, in which the carbon molecules are induced to arrange themselves automatically into either a soccer-ball-like configuration or a very thin cylinder when an electric current is passed between two graphite electrodes.[6] Carbon nanotubes are highly conductive of both electricity and heat and are one hundred times stronger than steel, giving them excellent potential in the design of tiny semiconducting circuits or energy-storing devices. They are currently mixed with other materials in the production of super-strong tennis rackets, lightweight bulletproof vests, and computer screens.[7] Researchers at the University of California have used carbon nanotubes to construct nanoscale radio components.

A second way to produce materials with features at the nanometer scale is from the top down. In this approach, one begins with a larger piece of material and uses photolithography or etching to add nanoscale features. An example of the top-down approach is the production of silicon microchips with nanoscale-sized circuits etched into them. Miniaturized hard disks, with magnetization at a nanoscale level, are used for data storage in almost all personal computers and handheld devices, minimizing the space needed for the storage of large amounts of data.[8]

A variety of inert materials have been developed at the molecular level. These include metals, ceramics, coatings, polymers, colloids, and aerosols that have been produced through structured chemical processes. There are already more than six hundred new products on the market that involve nanomaterials, ranging from sunscreens to catalytic converters, stain-resistant pants to tennis balls. As a branch of materials science, nanotechnology already provides us with improved products and tools that, while not life changing, are certainly life enhancing. New water-resistant fabrics are woven with pores at the nanometer level. These allow water vapor to escape yet are too small for water droplets to penetrate. A similar internal coating keeps tennis balls from losing air. This technology might soon be applied to car tires, resulting in an improvement in tire inflation that could lead to a significant savings in fossil fuels and reduced pollution.

One of the largest current markets for nanomaterials is the cosmetic market. Titanium dioxide and zinc oxide are found in a variety of sunscreens and cosmetics. The particles used in their formulation are too small to reflect visible light, thus, are invisible while being large enough to block the shorter wavelengths of ultraviolet light. Other cosmetics tout the virtue of particles small enough to penetrate the cells of the outer layer of skin.

Pharmaceuticals also stand to gain from nanotechnology in a big way. Nanosilver appears widely in a variety of products, ranging from food storage containers to shoe liners, due to its antimicrobial properties. Even the ancient Greeks and Romans used silver as a topical antibiotic. Silver is toxic to germs, even viruses, because it bonds to various parts of the organism. Yet, in minute concentrations, it does not harm human cells. In 2006, the government of Hong Kong sprayed a nanosilver coating on the handrails of the city's subway system in hopes of slowing the spread of avian flu.

Research has recently moved from the realm of inert materials into the realm of structures that change properties while in use. This is particularly important in medical uses of nanotechnology. There is much current research into nanoscale polymers for drug capsules, which release drugs only under certain conditions, such as the presence of acid or heat. These could be used to deliver drugs to specific locations or to target particular tissues, such as tumors. Doctors at the University of Texas have been experimenting with gold-coated glass nanoshells, which enter tumors through the blood vessels that feed them and are then heated with a laser to burn away the gold, and with it, the malignancy.[9] Pharmaceutical companies are also investigating the reformulation of several drugs themselves into nanoscale particles, which might be better absorbed by the body. Nanoscale diagnostic chemicals, such as those used for contrast in magnetic resonance (MR) imagery, show promise for improved targeting of particular organs or cells; other forms are being developed that will last longer in the body, allowing for MR imaging of arteries and veins.[10]

The primary short-term medicinal use of nanotechnology may well be a fairly simple one—the development of inexpensive water filters. Diarrhea is the leading cause of death among children in the developing world. According to the World Health Organization, water-borne diseases cause the death of more than 2.2 million people per year.[11] Carbon nanotubes can be woven into a filter with nanosized pores that allow water particles to pass through but block larger contaminants, such as chemical pollutants and bacteria. Such filters could be used not only for purification but for desalinization, and they accommodate a fairly rapid flow, since nanotubes are straight as opposed to conventional fibers, whose bends and convolutions impede water flow. A second approach to water purification involves the use of nanoparticles to absorb arsenic and other large contaminants.

The extremely small size of nanoparticles, and their unpredictable nature, makes nanotechnology hard to regulate. Whereas over $1 billion has been allocated to nanotechnology research in the

Unites States each of the last few years, only $38 million of that goes to assessing the environmental or human risks, according to David Rejeski, director of the Project on Emerging Nanotechnology at the Woodrow Wilson Center. There is neither a federal agency nor any statutes to govern nanomaterials. In the U.S. they might be regulated by the Occupational Safety and Health Administration during production, then by the Food and Drug Administration during use, falling to the Environmental Protection Agency when they enter the disposal segment of their life cycle. Regulations, moreover, are hard to draw up for such tiny materials, which are hard to monitor outside the laboratory.[12]

Nano-unpredictability extends to health and chemical safety. If inhaled, the particles are more likely than other substances to breach the blood–brain barrier. Research has shown that some nanoparticles, including carbon nanotubes, can result in the kind of protein fibrillation that is characteristic of Alzheimer's and Parkinson's disease.[13] By their nature, nanoparticles are chemically unpredictable, having a "high potential for being surprised," says Rejeski.[14]

At the scale of a nanoparticles, some elements behave in new ways, causing unusual "quantum" effects. For example, metals such as copper or zinc become transparent. Other metals become combustible. At the nanoscale, gold turns liquid at room temperature and becomes a catalyst, rather than an inert substance. Silver becomes antimicrobial. For now, however, nanotechnology seems to be under the control of the usual safety procedures. So far, it has not produced anything that utterly changes our world. But to hear Eric Drexler, that time may come soon.

The Philosopher's Stone

Drexler, one of the earliest visionaries of nanotechnology, explored the notion of molecular manufacturing in his 1986 book *Engines of Creation*. He also cofounded the Foresight Institute, which is pushing for guidelines to avoid mishaps with nanotechnology. Drexler worries about what he calls the "gray goo problem," the risk of fundamentally altering the chemistry of the world. Scientist Bill Joy notes: "Gray goo would surely be a depressing ending to our human adventure on Earth, far worse than mere fire or ice, and one that could stem from a simple laboratory accident."[15] Melting the world into goo, Drexler suggests, might not be the science fiction it seems at a time when rogues are looking for weapons of mass destruction.[16]

Nevertheless, Drexler takes us down the pro-nano path of Feyman's original vision. With the assistance of molecule-sized nanobots, Drexler says, we could construct material products one atom at a time. These machines could manufacture "anything that the laws of nature allow to exist." Indeed, we finally could turn base metals into gold. Or into fresh food, fresh air, even items that have no static nature, such as "clothing that becomes your bath water and then your bed." Nanosized robots could swarm through the bloodstream, repairing damaged cells, adjusting hair or skin color, and restoring lost youth.[17]

Could the dreams of the old alchemists finally become reality? Adam Keiper, writing in *The New Atlantis,* notes that so far no solid argument has been advanced that proves these things could not be done, though a number of scientists, including Nobel Prize–winner Richard Smalley, dismisses Drexler's vision as "just a dream," one that will "always remain a fantasy."[18] While no one yet knows how to make the molecular machines that could make these dreams a reality, neither has anyone found a fatal flaw in Drexler's theory. Dr. Bill Hurlbut describes nanotechnology as an "enabling technology" that "clearly is going to the bottom of the powers of matter, fundamental forces of nature to serve that which our human nature thinks nature ought to be, including revisions, potential revisions, of human nature" itself.[19]

Inventor Ray Kurzweil espouses Drexler's dream, particularly in its potential to change human nature. He expects that nanotechnology will be one among many technologies (such as artificial intelligence, robotics, and genetic engineering) that will irreversibly change human life. Kurzweil asks what will remain unequivocally human in a world in which physical reality is as mutable as virtual reality. His answer: "Ours is the species that inherently seeks to extend its physical and mental reach beyond current limitations."[20]

The twentieth-century theologian Reinhold Niebuhr spoke of this same human ambition, but he emphasized its perils. We are finite but strive to be infinite, Niebuhr, in *The Nature and Destiny of Man*, writes: "Man can find his true norm only in the character of God but is nevertheless a creature who cannot and must not aspire to be God."[21] This search to be like or to image the transcendent God, therefore, always includes a temptation to "sin," which Niebuhr defines as the "willful refusal to acknowledge the finite and determinate character of [human] existence."[22] If nanotechnology allows us to be godlike, then Niebuhr has already offered this warning against nano-hubris: "[Our] ability to stand outside and beyond the world tempts man to megalomania and persuades him to regard himself as the god around and about whom the universe centres. Yet he is too obviously involved in the flux and finiteness of nature to make such pretensions plausibly."[23]

Niebuhr wrote on the cusp of the revolution in atomic energy, but his principles surely apply to all technological advances. He urged human actions to solve problems, and the nanotechnology industry, by all reports, is moving ahead with safety and prudence in mind. But Niebuhr's point was also about the risk of taking wonder out of the world in our efforts to gain control. The wonder and meaning of life are in a source beyond human control and possession. He notes: "We can participate in the fulfillment of the meaning [of life] only if we do not seek too proudly to appropriate the meaning as our secure possession or to effect the fulfillment by our own power."[24]

When Martin Luther sat at his Wittenberg table discussing alchemy, he ended his comments by saying that changing one metal into another was an allegory for "the resurrection of the dead at the last day." Newton spent the last years of his life intensely studying alchemy and trying to decipher the timetables and secret numbers of the Bible in his own quest to come to terms with immortality. The quest for life everlasting continues in nanotechnology. Robert Freitas, a leading writer on nanomedicine, predicts that nanosized robots equipped to repair cells and deliver medications will give us the ability to halt aging and even reverse it:

> Once nanomachines are available, the ultimate dream of every healer, medicine man, and physician throughout recorded history will, at last, become a reality. Programmable and controllable microscale robots comprised of nanoscale parts fabricated to nanometer precision will allow medical doctors to execute curative and reconstructive procedures in the human body at the cellular and molecular levels. . . . [T]he ability to direct events in a controlled fashion at the cellular level is the key that will unlock the indefinite extension of human health and the expansion of human abilities.[25]

Today, hundreds of people have sought this indefinite extension of life by joining the extropian movement, which promises to preserve bodies into the future. Extropians have paid to have their bodies frozen in hopes of a nanotechnological resurrection at some future date. This is far from the resurrection promised in the Gospel, of course. The biblical resurrection is not about more time in our finite bodies. According to St. Paul, it is beyond time, a transcendence in which "we will all be changed" (1 Corinthians 15:51).

The old alchemy was actually far more mystical than we can imagine today. Its advocates wrote about how its secrets and formulas changed hearts, as if into gold. Today's nanotechnology also reminds us how fundamentally changeable our world is, since nothing is static or immutable. Our reality is formed of atoms that are in continual motion. They form one thing now, but they could, and will, easily form something else in the future. The "solid" world is ultimately quite malleable. Both the universe and the human soul are dynamic, and this can surely evoke wonder.

Genetically Modified Crops

They call it "golden rice," an apt name for the value it holds for feeding the world. In this form of alchemy, normal rice has been turned into a superlative foodstuff. No wonder Monsanto, one of the largest food corporations in the world, has taken up the cause of genetically modified crops, shipping them to every corner of the planet. The saga of genetically modified (GM) crops is also a story of how commercial exploits may have unintended consequences, especially in developing countries.

As ever, genetically modified food reveals both the blessing and the curse of a new technology. We have improved foods, making them stronger and more nutritional, but by the act of "technology transfer," they are altering native ecosystems and radically shifting the economic balance between nations. From Monsanto's point of view, the agricultural innovation and transfer are absolutely necessary. "Despite advances over the past century, many people cannot earn enough money to purchase food, and they do not have the resources to grow enough food to feed their families," the corporation says in its 2006 report. "The environment is struggling to compensate for the toll that human consumption takes on raw materials, energy, and agriculture. Finally, vast numbers of people have grown reliant on easily accessible but nutritionally poor foods."[26]

Technology has come to the rescue in the past. The Green Revolution of the 1960s and 1970s introduced new high-yield hybrid strains of most food staples, such as wheat, rice, and corn, as well as new fertilizers, herbicides, pesticides, and irrigation methods. Together, adoption of these technologies more than doubled worldwide grain production by the mid-1980s, not only keeping pace with population growth but actually raising the amount of grain available per capita by 40 percent. In the last two decades agricultural production has continued to grow by more than 2 percent a year. Despite this growth, more than 820 million people remain malnourished and a child dies of hunger every four seconds, making hunger the number-one health risk worldwide.[27]

Many look toward genetic engineering as the technology that will initiate a second Green Revolution and allow global agriculture to keep pace with global population growth. A much-touted exemplar of this second agricultural revolution is golden rice. The World Health Organization ranks deficiencies of iron, vitamin A, and zinc among the top ten causes of death in developing nations.[28] Vitamin A deficiency causes blindness in 250,000 to 500,000 children each year. Rice plants naturally produce beta-carotene (which is converted to vitamin A in the human body), but only in the green photosynthesizing parts of the plant. In 1999, researchers Ingo Potrykus and Peter Beyer developed a type of rice that contains beta-carotene in the endosperm, the edible part of the grain that remains after polishing, by adding genes originally found in daffodils and a strain of bacteria. Potrykus and Beyer's rice contained minimal amounts of beta-carotene, only 1.6 micrograms per gram of rice; however a second strain, developed in the United Kingdom, contains up to twenty-three times that amount, making it possible for a child's daily intake of rice to provide up to half the daily requirement of vitamin A.[29]

The wheat, rice, and corn that we depend on today are plants that do not and, in fact, could not exist in the wild but have been developed over the centuries through selective cultivation and breeding. Individual plants with naturally occurring genetic variations providing a desired characteristic would be selected to breed subsequent generations. The difference between a genetically modified crop and an ordinary hybrid is that modern methods of gene identification and splicing allow for the transfer of genes that do not naturally occur in a species, often introducing genes from an entirely unrelated species. Under normal conditions, rice would not cross-pollinate with daffodils. The addition of bacterial genes into plants goes even further, crossing from one type of organism to another. Genetic engineering not only allows the introduction of genes normally not available through breeding, but also allows for a new precision. One gene may be altered in a plant that contains thirty thousand genes, whereas in traditional cross-breeding a change in one gene may bring with it a large number of unintentional collateral changes.

Although golden rice is for the GM revolution what a movie star is for Hollywood, it is actually not yet commercially available. However, other GM crops have been widely adopted, especially in the United States.[30] In 2008, the United States approved more than seventy genetically modified crops; and the U.S. Department of Agriculture reported that 86 percent of the cotton, 80 percent of the corn, and 92 percent of the soybeans planted in the United States were genetically engineered.[31] These crops have been engineered not for a different nutrient content but for pest and herbicide resistance. GM cotton and corn are engineered to be pest resistant through the introduction of a gene from a soil bacterium, Bacillus thuringiensis (Bt), which causes the plant to produce a protein that is toxic to certain pests, including the European corn borer and the American cotton bollworm. Bt seeds produce a hardy plant that needs minimal pesticide application. On the herbicide front, Roundup Ready soybeans contain a gene that makes the plant resistant to Roundup Ultra, the trade name for glyphosate, a popular broad-leaf herbicide. This allows farmers to spray their fields, killing everything but the soybeans. While Roundup Ready soybeans produce slightly lower yields than regular seeds, farmers save on labor costs since less tilling is required to control weeds.

Globally, adoption of GM crops has been much slower. Tomato puree from genetically modified tomatoes was introduced in Europe in 1995 to initial support, but the inclusion of Roundup Ready soybeans in a variety of unlabeled products, coupled with other food scares, such as the outbreak of mad-cow disease in 1996, led to a green backlash. There is now virtually no market for GM foods in Europe. Nonfood products have fared a bit better. In the European Union (EU), genetically engineered rapeseed, used primarily for biodiesel and industrial lubricants, and corn grown for animal fodder, are the largest GM crops. While critics of GM crops in the EU have magnified their risk, even proponents recognize that there are potential problems. Most obviously, the introduction of new genes could generate new proteins in foods, proteins that could cause serious allergic reactions in people who have previously eaten these foods with no problem.

A far more pervasive concern is that of genetic drift. Since plants reproduce through the distribution of pollen, once a GM crop is planted outside the laboratory it is impossible to keep pollen from that crop from migrating to other plants and nearby fields. This is unlikely to cause a problem in the case of a crop like golden rice. However, most GM crops, such as Bt corn or cotton, are modified to be insect or herbicide resistant. Any spread of the Bt gene could speed the development of Bt-resistant pests; similarly, gene transfer from Roundup Ready soybeans to related weed species could make those weeds herbicide resistant and, hence, much harder to control.

With the U.S. market reaching saturation and the EU market remaining skeptical, agricultural corporations have turned to the developing world for both a secondary market and a better humanitarian image. Corporate reports, web sites, and advertisements cite the need for a second Green Revolution through genetic engineering in order to feed the growing population of the third world.[32] While GM crops do bring a potential for a second increase in global agricultural production, their adoption, particularly in developing countries, has been controversial. GM crops are paradigmatic of the problems that can result from globalization and, more particularly, from technology transfer from the first world to the third. They represent an example of a technology developed to suit a Western set of cultural and economic norms that cannot be simply transferred to a different cultural milieu.

India: The Problems of Technology Transfer

Because of its vast dependence on agriculture, India may be the poster child of the benefits and tragedies of the GM revolution. Of the twenty-two countries that cultivate GM crops, India has seen the fastest increase, since two-thirds of the population is engaged in farming. Yet India uses few GM foods to feed its population. The GM industry in India is dedicated to primarily one crop: cotton. This has made India the world's second-largest producer of cotton, which is 30 percent of its gross agricultural product.

The vast cotton fields of India, however, tell a tragic story. They make up the world's largest planted acreage of cotton but show the lowest yield in the world. The fields are battered by crop failure, unpredictable monsoons, and pest infestation. Bt cotton was introduced to India in 2002 in part to combat the latter. In the five years since, adoption has grown to 60 percent of all hybrid cotton planted. Yield gains over these years have averaged around 40 to 55 percent. Bt cotton has also resulted in a 39 percent reduction in pesticide use over regular hybrid varieties.[33]

These statistics are impressive. However, there is another equally impressive statistic. In the cotton-producing area of Vidarbha, more than five thousand farmers have killed themselves since 2002, an average of three per day in 2006.[34] Most of these suicides are committed by farmers who find themselves deeply in debt. Sixty percent of India's farmers own less than one hectare (2.5 acres) of land, and 95 percent report that their income does not exceed their expenditures.[35] Many farmers are only one failed crop away from economic disaster.[36]

While Bt cotton provides some insurance against one type of pest infestation, it does nothing to ensure against drought, flood, or other pests, and the seed costs nearly twice as much as ordinary hybrids. While India has price supports and subsidies for grain crops, policies initiated during the Green Revolution of the 1960s in order to help farmers buy hybrid seed and machinery, it has no direct subsidy for cotton.[37] Unaided, Indian farmers must compete on a global market with cotton-producing countries such as the United States and China, where cotton is highly subsidized. The high expense of Bt seeds also forces farmers to plant as much land as possible in a single crop, thus militating against the diversity that might ensure some income even when one crop fails.

Bt seeds raise ecological concerns for Indian farmers as well as economic ones. Pest exposure to Bt is much higher in a genetically modified crop than it would be were a pesticide (including Bt) applied topically. This increases the likelihood of the eventual development of resistance to Bt through natural selection in the insect population. In the United States, the Environmental Protection Agency has mandated that a certain percentage of a farmer's field be set aside for non-Bt crops in order to slow down resistance development. This is not expected in India. A secondary problem is that Bt remains active in crop residues, which poorer farmers routinely plow into the soil. This can reduce the fertility of the soil by producing toxins that reduce the activity of soil microorganisms. Thus, farmers who plant Bt crops risk the dual development of hardier predators and reduced soil fertility.[38]

GM crops also threaten to move the production of certain commodities away from the developing world to the first world by designing new crops to provide substances formerly grown only in the tropics. For example, rapeseed has been engineered to produce lauric acid, a component of soap that has previously been derived from coconut and palm oils. Thirty percent of the population of the Philippines is employed in the coconut industry.[39] Alternate, and possibly cheaper, sources of lauric acid would have serious ramifications for job security there. Similarly, the ability to produce vanilla or cocoa in temperate climate crops could be potentially devastating to the economies of Madagascar or West Africa.[40]

GM seeds are only one component in an increasing industrialization and Westernization of Indian farming. Harvard geneticist Richard Lewontin notes that India is undergoing a parallel change to that which occurred in the United States earlier in the twentieth century. The farmer in the first half of that century

> saved seed from the previous year's crop to plant, the plow and tillage machinery was pulled by mules fed on forage grown on the farm, 40 percent of planted acreage was in feed crops, and livestock produced manure to go back on the fields. Now the seed is purchased from Pioneer Hi-bred, the mules from John Deere, the feed from Exxon, and the manure from Terra. . . . The consequence of the growing dominance of industrial capital in agriculture for the classical "family farm" has been the progressive conversion of the independent farmer into an industrial employee.[41]

As of 2005, ten corporations controlled over half of the world's commercial seed sales; Monsanto controls production of 88 percent of the land acreage producing GM seeds. Ten corporations provide 80 percent of global agrochemicals. Many of the same companies exert influence on commodities markets as well. Five grain trading companies have controlled 75 percent of the world's cereal prices since 2000. Thus, a handful of corporations controls both the inputs to agricultural production and the outputs.[42] The creation of GM crops represents but the most recent step in this process of aggregating control of agriculture to a few major corporations. While agricultural industrialization affects farmers worldwide, it has greater consequences for farmers in the developing world, where social safety nets are negligible to nonexistent. For the poor cotton farmers of Vidarbha, governmental compensation for widows often seems to be the only option and the suicides continue.

Property Rights

India represents one other dilemma for developing countries in the age of GM crops and aggressive corporations—the defense of property rights and patents on native plants and naturally produced hybrid seeds. In September 1997, a Texas company, RiceTec, was granted a patent on basmati rice. While the patent is for a specific hybrid of Indian basmati designed to grow well in Texas, it also initially included the right to market under the basmati name and any crosses with its proprietary hybrid. The Indian government challenged the patent in 2000, claiming that basmati applies only to rice grown in India and Pakistan and that RiceTec's use of the name jeopardized India's annual basmati export market, worth $300 million. They were also concerned that the patent would cover traits of basmati rice such as grain length and smell that were intrinsic to Indian-grown basmati.

Fifteen of RiceTec's original twenty patent claims were later rejected or dropped, including RiceTec's right to the term *basmati* and claim to unique qualities of the grain.[43] India was successful in this dispute. However, according to Alex Wijeratna, international food rights campaign coordinator for ActionAid, "Litigation and similar actions favour the rich at the expense of the poor. Communities often do not have the money to put challenges together and are just not used to doing it. This clearly works to the benefit of the companies." He also notes that to mount a successful litigation against a patent, "You have to demonstrate that the knowledge was already there but very often this only exists in an oral form. If it is not written down then proving anything is far more difficult."[44] The RiceTec case has led the Indian Council of Agricultural Research to begin storing the DNA pattern of hundreds of crop varieties and indigenous plants in a gene bank in order to establish proof of origin of the genetic material, in the event of future patent disputes.

The basmati dispute is one example of a global movement toward the privatization of foods as a corporate commodity. Corporations such as Monsanto require farmers to sign agreements that they will not save seed, and Monsanto has taken legal action against more than one hundred farmers in the United States and Canada when patented plants have been found in their fields in greater quantity than could be accounted for by pollen drift.[45] One solution to the problem of seed piracy was the attempt in the late 1990s to develop plants that produce sterile seeds. These so-called terminator seeds raised such a public relations furor that the seeds were never marketed. Unfortunately, many farmers in the third world have come to believe that all GM seeds are terminator seeds, and the industry has done nothing to disabuse them of this fallacy.

Beyond the question of the commoditization of the necessity of food, one might ask whether genetic modification and patenting of life forms lead to a view of life itself as a commodity. U.S. patent law allows for patenting anything man-made or any process that is "novel, non-obvious, and useful."[46] However, patent law has always prohibited patenting a product of nature. GM crops are a new creation, not found in the natural world. They represent human creativity, a new product. Genes and genetic sequences have also been patented, falling under the category of formulae.[47] Both plants and genes occur in nature; genes characterize the essence of a life form by giving it its identity. To patent either seems to confer ownership of a part of nature itself to an individual or corporation. This goes against the traditions of much of the developing world where patterns of collective ownership and development are predominant.

Theology has had a good deal to say about ownership of the natural world, since it is presumed to have been created by God and is thus owned solely by its Creator. United Methodist Bishop Kenneth Carder notes, "One of the basic principles of our church is that life is a gift from God. The patenting of life forms reduces life to its marketability. Gone is the fundamental principle that life is a gift that ought to be shared and nurtured."[48] In May 1995, nearly two hundred religious leaders issued a "Joint Appeal against Human and Animal Patenting" expressing such concerns. On the other hand, theologian Ronald Cole-Turner notes, "Human beings may own individual animals and plants, and their components may be bought, sold, and used for food. That is, we may own these things as long as God's prior claim of ownership is acknowledged."[49] However, there is a difference between owning individuals and owning a species. As Leon Kass points out, "It is one thing to own a mule; it is another to own mule."[50] Finally, in reference to the developing world, religious leaders point out that patents not only establish a relationship between a person or corporation and an object, but a relationship among persons in reference to that object. Genes are a common heritage, and to claim a monopoly over their use in some way is to deny others access to a part of that heritage.

An Ambiguous Technology

New agricultural technologies are at their strongest in their promise to alleviate world hunger. But alleviating hunger is more a social than a technological problem. "About 320 million people go to bed hungry in India each night; not because there is not food, but because they cannot buy that food," writes Devinder Sharma of the Forum for Biotechnology and Food Security. "Between 2001 and 2003 India had a record grain surplus of wheat and rice amounting to 65 million tonnes. If you could have stacked the bags of grain one on top of another there would be enough bags to walk to the Moon and back; that was the quantity of food lying in India, and yet 320 million people went hungry."[51] Roland Lesseps, Jesuit scientist at the Kasisi Agricultural Training Centre and an outspoken critic of GM crops in Zambia, has noted the same problem: "The surest path toward elimination of hunger and malnutrition is to eliminate poverty and the unjust social structures that underlie it. These social and economic inequalities will not be remedied, but will only be made worse, by [GM] crops."[52]

GM crops hold the promise of increasing yields and thus of alleviating hunger, but only when they fit integrally into the local economic system. The massive adoption of GM seeds by U.S. farmers clearly indicates that they see these products as beneficial. Yet the risks for many farmers in developing countries have outweighed the benefits. GM seeds and the monoculture style of farming they promote have increased social and economic imbalances. The precipitous rise in farmer suicides in Vidarbha are a dramatic and tragic symptom of a technology transfer that is not working within India's current cultural and economic structures.

This does not mean that GM technology might not be important, indeed crucial, for India or other developing countries in the future. If global warming advances at the rates current computer models suggest, it could be vitally important to develop crops that are heat and drought tolerant.

Jacques Diouf, head of the UN Food and Agriculture Organization, has noted that India stands to lose up to 20 percent of its cereal production should temperatures rise 1 to 3 degrees Celsius.[53] A recent study has found that rising temperatures in the period between 1981 and 2002 have already decreased average global income from corn, wheat, and barley by $5 billion per year.[54] Most GM crops, currently engineered for insect and herbicide resistance, confront problems common in the United States and other temperate climates. To assist the most vulnerable regions of the planet where temperatures are likely to increase the fastest, the scientific community will need to change the focus of current research, a shift that may demand an increase in public, as opposed to corporate, funding if GM crops are truly to play a significant role in meeting the challenges of feeding the global population of the twenty-first century.

Energy Technology

One reason that Herman Melville's long and ponderous novel *Moby Dick* sold so well in the nineteenth century was its detailed section on the technology of whaling. The great whaling routes of the Atlantic, in fact, opened up the sea lanes and placed on the map such harbors as the Galapagos Islands, the famous setting for Charles Darwin to begin his thinking about biological evolution.

Whaling was big because whale oil was the revolutionary fuel of the nineteenth century, a primary way to generate light and replace wax and tallow.

The hundred-year boom in whale oil drove most whale species practically to extinction. Fortunately for the whales, a Canadian geologist, Abraham Gesner, discovered in 1849 how to distil bituminous tar into kerosene. Thus began the era of a new fuel, oil. But oil is used for far more than illumination. It fuels our cars and trucks, heats our homes, and powers our factories, and is a major component of many of our consumer goods. As Pulitzer Prize–winning author Daniel Yergin notes,

> Though the modern history of oil begins in the latter half of the nineteenth century, it is the twentieth century that has become completely transformed by the advent of petroleum. Today, we are so dependent on oil and oil is so embedded in our daily doings that we hardly stop to comprehend its pervasive significance.[55]

Unlike the still-to-be realized dreams of nanotechnology, the alchemy of oil has already changed our world immensely. The technology of oil has become so ubiquitous that it will be difficult to change, and yet like every technology, oil too must be assessed for both its usefulness and the consequences it brings to individuals and communities—and especially our environment.

Fossil Fuels

Ask any group of oil experts—and they will disagree on the following question: when will world oil production peak? That peak is the point at which the volume of oil we draw from the ground starts to diminish, finally limited by our technological capacity and by the oil reserves themselves.

The Energy Information Agency says global oil production will probably peak around 2037.[56] Others suggest that peak has already occurred or is imminent.[57] Production has already peaked in fifty oil-producing nations, including the United States in 1970 and Great Britain in the 1990s. The last major oil field to be newly discovered was in 1976. Fatih Birol, chief economist of the Paris-based International Energy Agency, is pessimistic: "We are of the opinion that the public isn't aware of the role of the decline rate of existing fields in the energy supply balance, and that this rate will accelerate in the future."[58] Whether production has peaked or will peak in thirty years, oil is a finite resource. Increased consumption coupled with an inevitable decline in production will have economic, political, and ethical ramifications.

At a National Energy Summit in 2001, Secretary of Energy Spencer Abraham noted, "America faces a major energy supply crisis over the next two decades. The failure to meet this challenge will threaten our nation's economic prosperity, compromise our national security, and literally alter the way we lead our lives."[59] While this is likely, increasing oil prices and decreased availability will have a larger initial impact on developing countries. As well as being necessary for most industrial development, oil is a critical part of modern agriculture as a component in fertilizer and as fuel for tractors, harvesters, and irrigation systems. As oil passes $120 a barrel, it becomes difficult for developing countries to afford its import. The lack of affordable oil disrupts agriculture and aggravates deforestation, as the local population reverts to earlier energy methods.[60] Expensive oil will block development in many of these countries, resulting in increased poverty and political instability.

One might think developing countries with their own oil reserves would fare better, but this has not proven to be the case. In general, oil has obstructed both democracy and economic growth due to the monopoly on its extraction by corrupt governments. No developing country outside the Middle East has developed a successful oil-based economy.[61] Countries such as Nigeria, Chad, Guatemala, Congo, Bangladesh, and Kazakhstan, where the World Bank has funded oil extraction projects, have experienced increased conflict, corruption, human rights violations, and environmental degradation.[62] Few oil dollars ever make it to the level of the general populace.[63]

While oil has destabilized the politics of countries in Africa and central Asia, it is in the Middle East that the presence of oil is having the greatest political effect. Sixty-six percent of remaining oil reserves is in the Arabian Gulf. In the last fifteen years, the United States has fought two wars in Iraq, which is thought to have the world's second largest oil reserve, after Saudi Arabia. While neither of these wars was ostensibly about oil, it is difficult not to surmise that oil was a factor. Saddam Hussein made lucrative oil deals with France, Russia, and China in the final years of his reign. These deals excluded the United States and United Kingdom; however UN sanctions kept France, Russia, and China from engaging in major oil extraction. The 2005 Iraqi constitution guarantees a role to foreign oil companies, an unusual move in the Middle East, where all other oil reserves are government monopolies.

Under the currently proposed Iraqi Oil Law, two-thirds of current oil fields, and all discovered in the future, would be open to control by foreign corporations. In a press conference in August 2005, President George W. Bush gave as one reason for the 2002 war the need to keep Iraq's oil fields out of the hands of terrorists. In this speech, he compared his resolve in the war with that of Franklin D. Roosevelt.[64] This was, perhaps, a more apt comparison than Bush intended. The year 2005 marked the sixtieth anniversary of the historic meeting between Roosevelt and the Saudi King Abdul-Aziz at which they arranged the U.S. purchase of Saudi oil in return for U.S. military aid and forged the special relationship between the two countries that remains today. The potential of increasing pressure on the Saudi monarchy by Islamic fundamentalists to disrupt this relationship provides another source of energy insecurity for the United States and a second reason to procure some influence over Iraqi oil.[65]

The modern story of coal is equally colorful and politically charged. In the days before the 2008 Olympics, the government of China, despite its energy crisis, had one run of good luck. The summer rains had washed away the mass of coal-generated smog that typically covers Beijing. The city looked pristine under the gaze of global media coverage. The government also shut down its coal-burning industry. Nevertheless, it still had to insert a computer-generated fireworks display into the official television feed because a clear Beijing sky is never guaranteed. Such is the story of energy use in a coal-dependent China.

Everywhere in the world, coal is the most immediate alternative to oil. Reserves of coal are greater than reserves of oil, and its use as an energy technology is not new. Spurred by high oil prices and concerns for energy autonomy, coal use grew worldwide by 22 percent in 2006. Much of this growth has been in the developing economies of China, India, and Russia; however, the United States and Europe are building increasing numbers of coal-fired power plants as well. In early 2007, it was estimated that coal-fired generators were being brought on-line at the rate of two per week and that thirty-seven nations are planning an increase in their coal-fired energy production in the next five years.[66] There are serious environmental costs to replacing oil with coal.

Coal is a leading cause of both smog and acid rain. Coal contains sulfur, which turns to sulfur dioxide when burned. Mixed with moisture, this precipitates out of the atmosphere in the form of sulfuric acid. The building of dozens of coal-fired power plants in China in each of the last few years has led to over one-third of China experiencing serious acid rain; particulates and greenhouse gases from these plants seem also to be a major source of pollution in Canada and the American Northwest. Acid rain destroys forests, kills fish and other aquatic organisms, and contributes to health problems. In Norway, fish populations have disappeared from thousands of lakes. In central Europe, whole forests have been wiped out. A similar impact on the environment could have devastating effects on the dense population of China. Most smokestacks in North America have been outfitted with scrubbers to remove much of the sulfur; this technology will be required on all new plants in China beginning in 2008.

Current scrubbing technology does not address a second and perhaps more lethal byproduct from burning coal—carbon dioxide. The year 2006, with a growth of 22 percent in coal burning, also saw a record 3 percent rise in carbon dioxide emissions into the atmosphere. Carbon dioxide (CO_2) is one of the main gases implicated in global warming. While all fossil fuels emit carbon dioxide, coal is the worst, emitting 20 percent more CO_2 per unit of energy than oil. The burning of coal is responsible for one-third of the CO_2 added to the atmosphere each year—up to 1 billion tons.[67] According to the International Energy Agency, CO_2 emissions from all our fossil fuels could grow by anywhere from 55 to 90 percent by 2030.[68]

There is no longer any scientific dispute that the earth is warming up. The Intergovernmental Panel on Climate Change (IPCC) now states with "very high confidence" that this warming is in part caused by human agency. There is a direct correlation between the amount of carbon dioxide in the atmosphere and rises in global temperature. Although there are normal fluctuations in temperature from year to year, eleven of the last twelve years are among the twelve hottest years recorded. The ice sheets on Greenland and the Antarctic are melting, and glaciers in both hemispheres are receding. This is adding to an overall rise in sea level, a rise that is threatening the existence of some of the islands in the South Pacific. Tuvalu, Kiribati, Vanuatu, the Marshall Islands, the Cook Islands, Fiji, and the Solomon Islands are all threatened with inundation within the next forty to fifty years. Meanwhile, both droughts and storms are becoming more severe on all continents.[69] The IPCC projects that by the end of the century, the earth's surface temperature will be 5 to 10 degrees F warmer than it was at the beginning of the century.[70] Such a warming could have disastrous results for humankind. Ice in the Arctic would disappear in the summer and the Greenland ice cap could melt completely.[71] This would result in a sea level rise of from 4 to 8 meters, inundating many coastal cities, most of Bangladesh, the Maldives, and much of Florida, the Texas coast, the outer banks of the Carolinas, and the eastern half of Long Island.

Nuclear Power

Faced with the negative impacts of both oil and coal, some have called for a new renaissance in nuclear energy, especially since it produces no CO_2. In a June 2005 speech, President George W. Bush stated, "It is time for this country to start building nuclear power plants again. Nuclear power is one of America's safest sources of energy . . . without producing a single pound of air pollution and greenhouse gases."[72] The public does not agree that nuclear power is one of the safest sources of energy. The accidents at Three Mile Island in 1979 and Chernobyl in 1986 sensitized the world's population to the risks associated with nuclear reactors. Both accidents were the result of human error, for which there is no technological fix. Thus, no new reactor has been built in the United States since the year before Three Mile Island, partly a result of the "not in my backyard" syndrome. But the local populace is not the only one skittish of nuclear power. Peter Bradford, of the Nuclear Regulatory Commission, noted, "The abiding lesson that Three Mile Island taught Wall Street was that a group of NRC-licensed reactor operators, as good as any others, could turn a $2 billion asset into a $1 billion cleanup job in about 90 minutes."[73]

Nuclear power also carries the unresolved problem of reactive waste. According to the U.S. Department of Energy, in 2008, there was fifty-six thousand metric tons of spent fuel from nuclear reactors in the United States waiting for disposal.[74] It takes one hundred thousand years for plutonium to lose its radioactivity. Spent nuclear cells must be stored in repositories away from groundwater, earthquake, political unrest, or climactic change. Ideal storage spots may be impossible to find—especially a single spot for all nuclear waste. Although the Yucca Mountain storage facility in Nevada has the federal go-ahead to become *the* national dumpsite, that decision still faces heavy political winds, and its future is yet to be determined. There are currently several smaller disposal sites around the country, typically close to the various reactors. Opening such sites always raises issues of justice and public confidence. Nearby residents face a great risk for little benefit in return. They also have little confidence in government agencies that have routinely conducted their operations, particularly nuclear testing, in great secrecy and with little thought about protecting the local populace.[75]

Beyond the storage issue, radioactive waste could also be used by terrorists to produce a dirty bomb. There is no clear line between peaceful and military use of nuclear power. This has become clear in the recent dispute between the United States and Iran regarding Iran's determination to acquire nuclear capabilities. The Iranian government insists its intentions are purely for energy production; the U.S. government is rightly skeptical of this. All of today's nuclear powers, with the possible exception of North Korea and the United States, initially claimed their development of the technology was purely for civilian use until they detonated a test weapon. As Hermann Scheer, of the German Parliament, points out, "If for no other reason than this, propagandizing a renaissance for nuclear energy is hair-raisingly irresponsible. At a minimum, the prerequisite for a country to use nuclear energy is stability in that state's domestic politics and international relations. In how many of the world's countries can these be guaranteed and permanently maintained?"[76]

New and Renewable Energy Sources

Two new sources of energy widely touted in political circles today are hydrogen and ethanol. While both these show promise as niche energy sources, neither is a magic bullet that will replace fossil fuels as simply as kerosene replaced whale oil. In the case of hydrogen and ethanol, both consume a great deal of energy in the very process of making "renewable" and clean energy.

Once again, the policy enthusiasm typically starts at the top, at least rhetorically. In January 2003, President Bush announced a $1.2 billion initiative to develop hydrogen fuel cells as our major energy source. Hydrogen is a plentiful gas in our universe and produces neither particulates nor greenhouse gases when it is burned. The problem is that hydrogen does not occur by itself in nature. It must be separated out of larger molecular combinations, such as with carbon, as found in coal, oil, or biomass, or oxygen, as found in water. This gives rise to two problems. First, hydrogen must be produced by separating it from its natural source. This requires energy in the form of electricity. Thus, hydrogen is not a primary energy source, but a secondary one, more a way of storing energy than producing it. However, as a storage medium, hydrogen has a related set of problems. Since it never appears by itself in nature, once separated, hydrogen is extremely unstable. It wants to recombine with other molecules. For storage, it must be liquefied or compressed and stored in pressurized containers. Liquefaction occurs at -253C, another process requiring energy. Thus, while hydrogen might eventually provide an efficient energy carrier for automobiles or other vehicles, it is not an energy source.

Ethanol, an alcohol produced from corn, grass, or sugarcane, is one of the fastest-growing commodities today. It has the advantage of fitting with current fossil fuel technology; most cars on the road today can drive on a mixture of gasoline with up to 10 percent ethanol. The United States produces 4.5 billion gallons of ethanol a year, and this number is rising rapidly. Most ethanol in the United States is distilled from corn. Brazil distills ethanol from sugarcane, which supplies 18 percent of its domestic fuel consumption. The most efficient crop for ethanol production is probably switch grass, and the United States may eventually turn to this from corn. Ethanol produced from crops has the advantage of being a renewable resource. It is not a pure energy producer, however. It takes energy to plant the crops, sustain them, harvest them, and then distill them into ethanol. However, a recent survey of studies has shown that, unlike hydrogen, ethanol is "energy positive" in that it produces more energy than is spent in its production.[77]

The drawbacks to ethanol are environmental and economic. First, ethanol production is extremely water intensive. In 2006, the production of 676 million gallons of ethanol at Nebraska's fifteen plants used 2 billion gallons of water.[78] Large-scale ethanol production could lead to conflict over water rights, as water itself becomes a scarcer global commodity. Second, the demand for ethanol has more than doubled the price of corn on the commodities market in the United States. This has had repercussions in the food market, even beyond U.S. borders. During the winter of 2007, Mexico was forced to put governmental limits on the price of tortillas, which had risen by more than 30 percent the previous fall.[79] Food riots took place in Haiti. The price of other grains, such as wheat and barley, also rose precipitously as farmers switched to corn. The biggest problem with ethanol, however, is that it will never be a primary energy source. It is simply impossible to grow enough crops to meet the world's energy demands. As agriculture professor Darren Hudson notes, "If we planted fence row to fence row—the entire U.S.—we couldn't produce enough ethanol to supply consumption."[80]

In contrast to hydrogen and ethanol, free sources of power can be found in sun, wind, water, and the thermal energy of the earth itself. These are the truly "renewable" energy sources. Unfortunately, they are not receiving the investment they need to be developed soon. They currently supply about one-tenth of the world's energy. The majority of this comes from hydroelectric power plants that harness the cascading water of rivers or reservoirs. In the United States, wind turbines provide 11,000 megawatts (mw) of power, while hydroelectric plants provide 80,000mw.[81] Geothermal lags far behind, producing only 1 percent of the nation's power, and there is little hope for more soon since the Bush administration cut all funding for hydro and geothermal power development in 2007.

Advocates of renewable energy reject the typical governmental or industry arguments that it is more costly than conventional approaches. Traditional energy sources have not been inexpensive—they have simply been massively subsidized. In 2001, world energy subsidies were $244 billion, of which $53 billion was for coal, $52 billion for oil, $46 billion for natural gas, $48 billion for electricity, $16 billion for nuclear energy, and only $9 billion for renewable sources (the other $20 billion is likely for "other" such as infrastructure or administration).[82] At the same time, conventional energy seems inexpensive because the long-term costs—such as nuclear waste storage or carbon dioxide emissions—are rarely considered in the overall price tag. A carbon tax has been proposed as a way to bring fossil fuels closer to their true cost. While small carbon taxes have been levied in several countries in the European Union, such a tax is unlikely to gain much ground in the current U.S. political climate.

Solar power is perhaps the most equitable of all power sources. The sun shines everywhere. And, unlike wind, hydro, and geothermal energy, one can harness this power "off the grid." In other words, solar power is widely distributable; it does not require a massive up-front investment in a centralized power facility. Herein lies the problem with solar power. Once photovoltaic cells are installed, the energy to run them, sunlight, is free. It is a fuel source that is not controllable by the traditional energy corporations and, hence, not a lucrative investment.

After initial excitement in the 1970s, solar power has languished in the United States. The energy corporations have fought the introduction of solar technology in a variety of ways. During the energy crisis of the early 1970s, the Ford Foundation published a report entitled *A Time to Choose* that highlighted the need for renewable energy and the risks associated with nuclear power. The energy corporations countered with *No Time to Confuse* in which they sought to foster fears of renewable technology. They also bought up small solar companies and then quietly closed them. Governmental support for solar technology research and development was slashed under President Ronald Reagan and has yet to be reintroduced. Continued escalation of oil prices and concerns over global warming may lead to a renaissance in solar energy use and development in the near future. This will be long overdue.

When I was a child, we used to borrow a neighbor's tent trailer each summer for a week of camping in the Minnesota north woods. The trailer was a rather basic affair, with a top that folded down on both sides to reveal a pop-up tent. Each year, after our trip, my mother would spend a day or two meticulously cleaning the trailer and mending any small tears in the tent top. One year, pressed into helping with this task, I pointed out that the tears we were mending had already been there when we had gotten the trailer. Mom said, yes, she knew that, but a part of being a good neighbor was always to make sure you returned anything loaned to you in better condition than when you received it. According to biblical tradition, the earth is on loan from its creator God, yet our reluctance as a society to move from fossil fuels to renewable energy sources makes it highly unlikely that we will be able to pass the earth on to future generations in better condition than we received it. Frequently, concern for the environment has been pitted against concern for human welfare. This is a false dichotomy, for human welfare depends on a functional ecosystem.

There is little of our earth that has not already sustained human interference. As of 1995, only 17 percent of earth's land mass has escaped the impact of human technology.[83] There is no longer any place for the human family to go, should we render this place uninhabitable. I chose energy technology to be the last technology discussed in this book because, in some sense, it may be our last technology. In *Field Notes from a Catastrophe,* Elizabeth Kolbert writes, "It may seem impossible to imagine that a technologically advanced society could choose, in essence, to destroy itself, but that is what we are now in the process of doing."[84]

CHAPTER 5
Technology Goes Global

THE STORY OF our relations with our neighbors has been told since time immemorial. One of the most famous was told two thousand years ago. When a Jewish lawyer in Palestine asked Jesus of Nazareth, "And who is my neighbor?" he answered with the parable of the Good Samaritan. In this story, where the Samaritan helps a complete stranger who was beaten on the dangerous road between Jerusalem and Jericho, Jesus enlarged the young lawyer's sense of community. The neighborhood was no longer limited to fellow Jews. As we have seen in these pages, technology has enlarged our neighborhoods, too.

The advent of new communication, information, and transportation technologies means we now encounter one another not only on the road but in cyberspace and over the airwaves. We now live in a world in which communication and transactions happen on a planetary scale in real time.[1] Money moves from the halls of Wall Street to the trading floors of Singapore with the click of a mouse. A call from South Dakota to a business help desk may be answered in Bangalore. A protest in Belgrade appears on television screens from Buenos Aires to Tokyo as it occurs. Pollution drifts from China to the forests of British Columbia. Western-induced melting in the Arctic threatens the Maldives. The local is now global. The road to Jerusalem is everywhere. This capacity to be a global species, making every environment livable, seems to be in our genes, says the Sri Lankan scientist Carlo Fonseka. "Biological evolution has not fitted humankind to any specific environment," he says. "Roses cannot bloom in deserts and chimpanzees cannot thrive in the north pole [sic], but Homo Sapiens can. In other words, humankind has the capacity to survive in any environment, including even outer space."[2]

What is new about this capacity is our rapid application of it, especially over the last three decades. Only since the 1980s have we had the technology to enable corporations, economies, and media to function efficiently on a planetary scale in real time. Computers, the Internet, satellites, cell phones, jet airplanes, and container cargo transport have allowed the rise of global business and media networks. No part of the world is isolated, and no country can develop its economy or society alone.

In this final chapter, we will look at the implications of technology's going global—implications that are environmental, moral, and political. The world is responding to our rapidly spreading technologies, and we see this in our religious traditions in particular. New global relationships will become a challenge to us all but may also inspire a greater sense of responsibility and an expanded ability to be good Samaritans in an ever-larger neighborhood.

The globalization of technology and of our technologically based world economy is here to stay. It has brought tremendous economic advantages to much of the business community. The negative impact is also well documented. Increased globalization has produced widespread economic disparity between the rich and the poor, between industrialized and developing nations. There is also a genuine fear that the consumerist spirit that underlies our global economy will lead to faster destruction of the world's natural resources. These two outcomes together ignite political and social unrest, with the potential for global disaster. *New York Times* columnist Nicholas Kristof offers this bleak political assessment: "As we pump out greenhouse gases, most of the discussion focuses on direct consequences like rising seas or aggravated hurricanes. But the indirect social and political impact in poor countries may be even more far-reaching, including upheavals and civil wars."[3]

How real are these fears? While the economic condition of many has improved over the last three decades and the proportion of the poor has decreased in China, India, and Brazil as a result of globalization, this decrease is not seen in the majority of the world's countries, especially in sub-Saharan Africa and many countries that made up the former Soviet Union. And, though the proportion of poor in Asia has decreased, sheer numbers of the poor have increased due to growing populations. One person in five in this world still lives on less than $1 a day, and one in seven suffers chronic hunger.[4] The distribution of wealth is increasingly polarized. The global ratio of income between the top 20 percent and the bottom 20 percent was thirty to one in 1960. In 1994, it was seventy-eight to one. The top 385 billionaires hold personal assets larger than the GDP of all the countries representing the poorest 45 percent of the world's population put together.[5]

Despite these statistics, it is clear that a globalized economy has brought jobs to many in the developing world. However, Manuel Castells, in a report for the United Nations Research Institute for Social Development, notes two trends, furthered by the nature of globalized business, that make these jobs problematic. The first is a growing individualization of labor. There is little collective bargaining, and persons are generally hired on an individual basis, not as a family or village unit. This opens the door to economic exploitation and to a prevailing sense of insecurity among workers, an insecurity that leads many to agree to unacceptable levels of pay or working conditions.

There is a second trend toward social or economic exclusion. Here individuals, groups, or even entire nations find themselves left out of the global network. Consider the large numbers of women from the former Soviet Union and eastern Europe who have resorted to prostitution because they have no other option for work. The recent rise in piracy off the coast of Somalia demonstrates what can happen when an entire country, due to a lack of resources and the political unrest such a lack engenders, finds itself "off the grid."[6]

Changes in technology in one part of the world can also have significant unintended impacts on the lives of those in another place. In other words, local control over work, commodities, media, and the environment is being lost under globalization. For example, when the United States, Brazil, and the EU use corn to produce ethanol, they change the price and availability of food in places as distant as Mexico, Haiti, and Somalia. Genetically modified seeds have led to higher yields and greater prosperity for American farmers, yet their introduction in India has led to bankruptcy and suicide for many farmers there.

At one time, it was a somewhat local debate in a society over the creation and regulation of a new technology. However, globalization has put that beyond local reach. Consider the debate in the United States over using fetal stem cells for research. President George W. Bush imposed a federal ban on research using any but a small number of existing lines of stem cells. However, the effect of this ban was simply to move research overseas. Australia, Belgium, China, India, Israel, Japan, Singapore, South Korea, Sweden, and the United Kingdom all allow not only the use of fetal stem cells in research but also research into somatic cell transfer for therapeutic cloning. Brazil, Canada, France, Iran, South Africa, Spain, the Netherlands, and Taiwan are somewhat more restrictive but do allow research with fetal cell lines from embryos that would otherwise be discarded by fertility clinics. Each of these countries has multiple centers engaged in biogenetic research.[7] Exercising local control over the development of a new technology may thus be futile.

Similarly, global technology is difficult to control at the local level, as the penetration of the Internet has proved in authoritarian states such as Saudi Arabia, Myanmar, Cuba, Iran, or China. Myanmar has been, perhaps, the most successful in banning public access to the Internet and Western media. But this country represents an extreme. Iran bans the use of individual dishes for the reception of satellite television. Yet when I glanced out of my fourteenth-floor hotel window during a recent trip to Teheran, it was obvious that this is a hard ban to enforce: rooftops bristled with such dishes. China has, so far, been fairly effective in restricting Internet use. It blocks some web sites until suppliers restrict content, and it makes selective arrests to promote self-censorship. This has not prevented a growth of political discussion in Chinese chat rooms nor the introduction of new political and cultural ideas to an increasingly web-savvy young population.

In Saudi Arabia, it is estimated that two-thirds of the more than four hundred thousand Internet users are women. They find in this new medium an escape from strictly cloistered lives. A whole new practice, called "numbering," has sprung up, in which young men chase cars containing young women and attempt to give them their cell phone numbers. Many young Saudis of both sexes have pages on Facebook and increasingly communicate on-line, raising new questions regarding the strict segregation of men and women in that society.

The problems of Facebook pages in Saudi Arabia or of Iranians watching *South Park* or *Desperate Housewives* show that globalization is not just an economic phenomenon. Much of the fear of globalization and new technologies in the developing world stems from the cultural reach and appeal of both Western products and Western values. The ubiquity of McDonald's, Coca-Cola, Nike, and Levi's represents not just a marketing success but the allure of the American way of life, with its emphasis on youth, individualism, consumerism, and materialism. The dark side of America is also exported, including the violence endemic to American movies and video games and the exploitative sexuality of pornography. The globalization of culture appears to be pushing local customs and norms into an increasingly niche-like position, turning the traditions of centuries into quaint customs, if not into history. This has provoked a reaction—most notably among strict religious traditions.

Religious Responses to Globalization

In a global world, religious traditions are responding to technology in different ways. The responses are conditioned by both the religion's underlying understanding of the nature of humanity, divinity, and the world and by the more particular codes of conduct that each religion propagates. The worldview espoused by a particular religion can influence a society's enthusiasm for developing and/or using a new technology.

Japan, a leader in the field of robotics, is a revealing case. The Japanese are particularly fond of human-like robots. Part of this comes from the animist spirit of the Shinto religion. For a Shinto, objects in the natural world are imbued with spirit. Rocks, trees, and springs all have a spirit of their own. There is no sharp distinction between the ensouled and nonensouled. In this world, an animate computer is quite acceptable as having something like the emotions of a living creature. In Japan, robots feed the ill in hospitals, vacuum corridors, comfort children, work as receptionists, and even serve tea. They are often welcomed at their first day of work with a Shinto ceremony. Robots are portrayed positively in Japanese art and science fiction. "Japanese thought stresses harmony and does not tend to see confrontational situations," says Masahiro Mori, professor at the Tokyo Institute of Technology. "The idea that robots may be a potential enemy just doesn't exist in Japan."[8]

On the other hand, religious codes of conduct may also make it harder for a particular society to accept or use a new technology. Among Christians, concerns over the propagation of genetically modified crops center on issues of social justice: whether they damage farmers in the developing world or really do solve the problem of hunger. They do not ask whether GM foods are proper to eat, for Christians have no dietary laws. But for Jews and Muslims, the question of whether foods containing genetically modified ingredients are kosher or *halal* must be asked. Neither community has reached a consensus on this issue.

The major Orthodox Jewish authority, the Orthodox Union, considers these foods to be kosher so long as nonkosher genes are not implanted directly into the food source. "Nonkosher genes serve only as a chemical template," the union says. "The template is then reproduced onto materials taken from yeast which are then introduced into plants via bacteria. The reproduced gene in the plant is therefore from a kosher source."[9] In other words, what is ultimately introduced into the plant is itself plant material, taken from yeast. Thus, this form of genetic modification is not substantially different from selective breeding, where material from one plant is deliberately introduced into another. However, others disagree. Writer Michael Greene points to the prohibition in the Torah against the mixed breeding of crops or domestic animals: "You shall not let your cattle mate with a diverse kind; you shall not sow your field with two kinds of seed" (Leviticus 19:19); "You shall not sow your vineyard with two kinds of seed" (Deuteronomy 22:9–11). He takes his premise from a long tradition of Jewish thinkers, including the historian Josephus, who writes, "Nature does not rejoice in the union of things that are not in their nature alike." Islam faces the same quandary in determining whether genetically modified foods are *halal*. While the Council of Ulemas in Indonesia answers this question in the affirmative, the U.K. Islamic Medical Association considers genetically modified foods "against the basic Islamic teachings, against God and the Qur'an . . . These are all the beliefs of 1,400 million Muslims in the world."[10]

Another source of religious resistance to global technology is a concern over its impact on families and community life. Technology has changed the nature of family life in the Western world. The introduction of labor-saving devices such as dishwashers, washing machines, vacuum cleaners, and microwaves has freed women from the necessity of spending most of their hours caring for the home and family. On the other hand, these things cost money, and their cost, along with that of the entertainment technology most families now find indispensible, has been one factor that has propelled women out of the home and into the workplace. The traditional family, in which the man was the breadwinner, was an economic unit in which the woman was inherently at a disadvantage. Her sphere of influence was confined to the home and family, and divorce was uncommon, partly because it was not economically viable for the woman. Technology has changed this. It has led to what British sociologist Anthony Giddens calls a democratization of relationship in which men and woman participate on a much more equal level. While he believes this sort of relationship brings far greater happiness and fulfillment to more people, it is viewed as a threat by those who see the increasing equality of women as undermining notions of male authority, female virtue, rearing children, and the sanctity of the home.[11]

Finally, globalization can be seen as a threat to the internal truth claims of a religion. Never have so many believers been in touch with people from other religious traditions. The tenets of a variety of faiths are available at the click of a mouse. Global mobility sets up a world where one's neighbors may come from a wide array of religious backgrounds. Harvard professor Diana Eck, in noting that the United States is the most religiously diverse country in the world, says this diversity goes much further than the mere presence of a variety of faiths. It is not just that there are Buddhists in America: there are Sri Lankan, Tibetan, Thai, Burmese, Chinese, Taiwanese, Vietnamese, Laotian, Cambodian, Korean, and Japanese Buddhists, as well as some homegrown versions of our own. Each community has its own ceremonies, books, and places of worship.[12] Eastern Orthodoxy in the United States is represented by immigrants from Russia, Ukraine, Greece, Serbia, Macedonia, Lebanon, Palestine, Armenia, Eritrea, Syria, and Egypt. Again, each has its own rites and ceremonies, yet, lacking a large immigrant community, many find that they must worship together in a new amalgamation of Orthodoxies. In parts of Europe, where communities used to be homogenously Protestant or Catholic, Islam is growing rapidly and various Eastern faiths entice the youth. Cultures shift and blend, as I observed recently in a park in Ljubljana, Slovenia, where I was serenaded by a band of accordion-playing Hare Krishnas.

Contact with so many religious faiths has led to both increased conversion of individuals from one faith to another and to the fairly new phenomenon of multiple religious belonging. A Pew Forum on Religion and Public Life report, "U.S. Religious Landscape Survey," depicts a highly fluid and diverse national religious life. If shifts among Protestant denominations are included, then it appears that 44 percent of Americans have switched religious affiliations.[13] This fluidity also finds expression in the number of individuals who practice more than one religious tradition. This has long been an accepted practice in Asia. A Japanese woman once told me, "We are born Shinto, we marry Christian style, and we are buried Buddhist." But this is a relatively new phenomenon among adherents to Western monotheist religions. It is no longer unusual for a Lutheran parishioner to go to services on Sunday while also attending a meditation group on Fridays, or for a Catholic avidly to be reading the poetry of Rumi. Many, while staying loyal to the faith in which they were raised, find inspiration and even a new or clearer religious understanding in a secondary tradition. I, myself, spent some of my happiest years as a graduate student attending Quaker meetings on Sunday mornings and serving as cantor for the Russian Orthodox on Friday evenings.

One of the pioneer scholars of interfaith dialogue, William Cantwell Smith, has described for us the implications of our new interreligious neighborhood, in which everyone has access to people and resources from a variety of religious traditions, access that has broadened considerably with the advent of the Internet.[14] This multi-religious neighborhood is not a comfortable place for many. Wesley Ariarajah, who heads the World Council of Churches' interfaith unit, explains why this discomfort arises: "In a global age where all other identities are under attack, religious identity, for those who have them [sic] explicitly or in the form of its cultural manifestation, has become the only identity that still gives them the sense of belonging."[15]

Increased contact with other religious traditions and regular contact with persons from other faiths have also meant that religious believers are now much more frequently called on to defend the truth-claims of their own faith. On the one hand, this has led to a greater understanding and interest in religion, per se, especially among the young, but it has also led to defensive posturing and the rise of fundamentalism in a variety of faiths. Fundamentalism is a relatively new phenomenon, the term appearing only around the turn of the last century, to describe a new movement among Protestant Christians to return to the "fundamentals" of the faith. As Giddens explains, fundamentalism is "beleaguered tradition":

> Fundamentalism isn't about what people believe but, like tradition more generally, about why they believe it and how they justify it. It isn't confined to religion. The Chinese Red Guards, with their devotion to Mao's Little Red Book, were surely fundamentalists. Nor is fundamentalism primarily about the resistance of more traditional cultures to Westernization—a rejection of Western decadence. Fundamentalism can develop on the soil of traditions of all sorts. It has no time for ambiguity, multiple interpretation or multiple identity—it is a refusal of dialogue in a world whose peace and continuity depend on it. Fundamentalism is a child of globalization.[16]

While the recent rise in fundamentalism can be seen as a reaction to globalization and technology, a search for the unchangeable in a changing world, fundamentalist groups have made extensive use of technology itself. There is a vast array of fundamentalist web sites, and evangelical preachers in the United States have exploited the opportunity to reach large audiences through television and radio. The terrorist attacks on 9/11 are an extreme example of the fundamentalist embrace of modern technology. Mohammad Atta and his group communicated extensively via the Internet, trained on sophisticated video simulations, and finally turned a technology of modern transportation into a weapon of mass destruction. In an odd, and sad, paradox, they could never have carried out their attack on globalization and the increasing hegemony of the Western world without the technologies that that world provided.

Relationship and Responsibility

In a world where technology is a global phenomenon, effective responses to technology cannot be local. Religious communities provide a locus for examining technology that is global in both time and space. In terms of time, religion provides a perspective that brings the wisdom of our forebears, through sacred texts and tradition, into dialogue with technology. The global nature of our religious communities also provides a wide perspective from which to evaluate technology. World religions speak for all their adherents, not just for a local community.

The nature of sin has changed in our globalized world. Theologian Alfred Schutze notes,

> Whereas only a few centuries ago evil, so-called, had to be considered pertinent to moral behavior, more specifically the backsliding or weakness of the individual, today it also appears in a manner detached from the individual. It shows up impersonally in arrangements and conditions of social, industrial, technical and general life which, admittedly, are created and tolerated by man. It appears anonymously as injustice, or hardship in an interpersonal realm where nobody seems directly liable or responsible.[17]

One way to understand this new stumbling block is to consider, as we have before in this book, how technology has distanced us from the effects of our actions. Philosopher Emmanuel Lévinas underlines the importance of face-to-face encounter in our post-modern world: "The relation to the face is straight-away ethical. The face is what one cannot kill."[18] A face makes a person real and immediate. The challenge, Lévinas says, is to extend our natural response to the face we know to people we shall never meet, in effect, neighbors in the abstract.

An off-Broadway play of the 1970s, titled *The Rescue,* dramatizes, using a lavishly romantic plot, this tension in us between responding to faces but being indifferent to people we put at a distance, often by way of technology. The story goes like this: during the Vietnam war, a young American bomber pilot is shot down. He flees his burning plane and is taken in by a farmer and his daughter. They care for his injuries, and, naturally, he falls in love with the girl. After several months, the Green Berets locate the pilot. As the Berets head out from the "rescue," the commander says that the farmer and his daughter are a risk to their escape and orders them shot. The pilot's plea for mercy fails to reverse the decision. So he insists on pulling the trigger himself. He goes into the thatch-roofed house, taking the family into a room, where he offers a tearful goodbye. Then he shoots himself.

With that intense ending, the playwright reveals a human paradox. As a bomber pilot, the young airman dropped explosives on large populations, feeling little compunction. They were statistics rather than real persons. The war plane's technology distanced him from his targets. Once he was face to face with a real villager, it was impossible to harm her.

In his public addresses, Andrew Kimbrell, executive director of the Center for Food Safety, often cites *The Rescue* to illustrate the problem of "cold evil," the evil not of anger or hatred but of distance and disinterest. This cold evil is not only expressed in the use of modern weapons but also in the global extension of technology. Kimbrell writes:

> Obviously, few of us relish the thought that our automobile is causing pollution and global warming or laugh fiendishly because refrigerants in our air conditioners are depleting the ozone layer. I have been in many corporate law firms and boardrooms and have yet to see any "high fives" or hear shouts of satisfaction at the deaths, injuries, or crimes against nature these organizations often perpetrate. . . . We are confronted with an ethical enigma; far from the simple idea of evil we harbored in the past, we now have an evil that apparently does not require evil people to purvey it.[19]

Putting the label of "evil" on our technological isolation from our neighbors—an isolation promoted by our cars, computers, phones, iPods, and BlackBerrys—is a very hard pill to swallow. The story of the Good Samaritan, however, can be just as demanding on us; we need not be the one who beat the man and left him on the road to be complicit in his plight. The neighborhood has gotten bigger, and some of our most destructive actions are not taken as individuals but as communities.

In effect, globalization and technology have altered our historical definition of sin. In an interview with the Vatican's daily newspaper, *L'Osservatore Romano*, church official Monsignor Gianfranco Girotti explained how the definition has changed. In the past, Girotti said, sin was considered an individual trait or activity, as seen in the traditional list of the seven deadly sins (anger, pride, lust, gluttony, greed, sloth, and envy). Now, it is a communal trait that damages society as a whole. His new list of sins includes pollution, genetic manipulation, drug use, and economic disparity between the wealthy and the poor.[20] Enabled by technology, these sins destroy the social fabric on a global scale. The World Council of Churches, which represents 348 Protestant and Orthodox denominations in more than 120 countries, has also urged a new sense of responsibility and stewardship on issues such as human genetic research, biotechnology and agriculture, climate change, and the economics of globalization.

Monsignor Girotti was not the first to use the seven deadly sins as a trope for the collective sins of our age. Mohandas Gandhi closed an article in *Young India* in 1925 with a list of seven social sins: politics without principles, wealth without work, pleasure without conscience, knowledge without character, commerce without morality, science without humanity, worship without sacrifice.[21] As India increasingly becomes a player in the global technological revolution, Hindu scholars and theologians note the same need for a communal ethic. Swami Jitatmananda writes, "Hinduism continuously asserts that happiness lies not in individualistic living, however excellent that may be, but in living a holistic life for the welfare of entire humanity, because each one of us is inextricably connected with the universe. 'I connect the whole universe like a thread connecting pearls,' says Sri Krishna in the Bhagavadgita."[22]

Yet communal life is, in the end, made up of the actions of myriad individuals. Gandhi's list of sins clearly ties action in the world to the inner life, the spiritual life of the individual. Muslim theologian Seyyed Hossein Nasr notes this connection:

> All these elements are tied together—new technologies, political systems, economic systems, and social structures—to affect the way things are changing. Knowledge brings power and the way to change things is also through knowledge. Those who have the correct knowledge of what is going on, who also practice what they preach, can bring about positive change.[23]

Nasr calls for a return to a clearer understanding of humankind's position in relationship to the world as set out in Genesis 1 and Surah 2 of the Qur'an:

> Having forgotten their vice regency today men are trying to act as gods, and they will be punished in the most severe way for this sin. I have always said that however powerful we may appear to be as we try to destroy nature, nature will have the final say. Nature has direct contact with God; it is not responsible to us. It is we who are responsible for its protection, because of the function that God has given us. He has given us intelligence, free will, and other powers which we must use rightly, always remembering that we are His viceregents.[24]

Nasr brings us back to where we started in chapter one. Genesis 1 tells us that humankind was made in the image of God, a God who is a creator, who has dominion over the natural world, and who speaks in the plural, embodying relationship in God's very nature. Creativity, responsible dominion, and relationship thus lie at the center of our being as well, in a continual tension. When the technologies we create cause us to lose sight of one of these three poles, they become destructive. Computers can inhibit relationships; our energy and agricultural technologies harm the earth, over which we have dominion; our drug use threatens the nature of human creativity itself. Yet our technologies, when they consider all three of these poles, can heal the sick, help the suffering, and feed the hungry. By way of the simple "solar" flashlight, they can light a hut in the African veldt and allow a child to study at night.

In our technology, as in most of our pursuits, we succeed and we fail. Our technologies bring us tremendous advantages, a portion of control over the natural world. But they will never bring us total control. Reinhold Niebuhr reminds us that our mistakes are generally caused by overreaching, prompted by "the desire to find a way of completing human destiny which would keep man's end under his control and power." But Niebuhr also finds hope in the fact that "faith completes our ignorance without pretending to possess its certainties and knowledge; and . . . contrition mitigates our pride without destroying our hope."[25] Religion gives us a platform from which to evaluate our technologies, a voice to call for a change in direction, if needed, and a call to contrition when we fail. In our technologies, as in our daily lives, we live out the words of an old Benedictine, who, when asked what monks did in the monastery, replied, "We fall and get up, we fall and get up, we fall and get up again." Hopefully, we will recognize when we are falling and get up again before it is too late.

Appendix

BELOW ARE philosopher Jacques Ellul's "76 Reasonable Questions to Ask about Any Technology":[1]

Ecological

- What are its effects on the health of the planet and of the person?
- Does it preserve or destroy biodiversity?
- Does it preserve or reduce ecosystem integrity?
- What are its effects on the land?
- What are its effects on wildlife?
- How much and what kind of waste does it generate?
- Does it incorporate the principles of ecological design?
- Does it break the bond of renewal between humans and nature?
- Does it preserve or reduce cultural diversity?
- What is the totality of its effects, its "ecology"?

Social

- Does it serve community?
- Does it empower community members?
- How does it affect our perception of our needs?
- Is it consistent with the creation of a communal, human economy?
- What are its effects on relationships?
- Does it undermine conviviality?

- Does it undermine traditional forms of community?
- How does it affect our way of seeing and experiencing the world?
- Does it foster a diversity of forms of knowledge?
- Does it build on, or contribute to, the renewal of traditional forms of knowledge?
- Does it serve to commodify knowledge or relationships?
- To what extent does it redefine reality?
- Does it erase a sense of time and history?
- What is its potential to become addictive?

Practical

- What does it make?
- Who does it benefit?
- What is its purpose?
- Where was it produced?
- Where is it used?
- Where must it go when it is broken or obsolete?
- How expensive is it?
- Can it be repaired?
- By an ordinary person?

Moral

- What values does its use foster?
- What is gained by its use?
- What are its effects beyond its utility to the individual?
- What is lost in using it?
- What are its effects on the least advantaged in society?

Ethical

- How complicated is it?
- What does it allow us to ignore?
- To what extent does it distance agent from effect?

- Can we assume personal or communal responsibility for its effects?
- Can its effects be directly apprehended?
- What ancillary technologies does it require?
- What behavior might it make possible in the future?
- What other technologies might it make possible?
- Does it alter our sense of time and relationships in ways conducive to nihilism?

Vocational

- What is its impact on craft?
- Does it reduce, deaden, or enhance human creativity?
- Is it the least imposing technology available for the task?
- Does it replace, or does it aid human hands and human beings?
- Can it be responsive to organic circumstance?
- Does it depress or enhance the quality of goods?
- Does it depress or enhance the meaning of work?

Metaphysical

- What aspect of the inner self does it reflect?
- Does it express love?
- Does it express rage?
- What aspect of our past does it reflect?
- Does it reflect cyclical or linear thinking?

Political

- Does it concentrate or equalize power?
- Does it require or institute a knowledge elite?
- It is totalitarian?
- Does it require a bureaucracy for its perpetuation?
- What legal empowerments does it require?
- Does it undermine traditional moral authority?

- Does it require military defense?
- Does it enhance or serve military purposes?
- How does it affect warfare?
- Is it massifying?
- Is it consistent with the creation of a global economy?
- Does it empower transnational corporations?
- What kind of capital does it require?

Aesthetic

- Is it ugly?
- Does it cause ugliness?
- What noise does it make?
- What pace does it set?
- How does it affect the quality of life (as distinct from the standard of living)?

Notes

Preface

1. http://dl.getdropbox.com/u/421123/Hajime_Interior.pdf. Nicholas Carr, "Is Google Making Us Stupid?" *The Atlantic*, July/August 2008.

Chapter 1: Of Morals and Machines

1. Will Connors and Ralph Blumenthal, "Solar Flashlight Lets Africa's Sun Deliver the Luxury of Light to the Poorest Villages," *The New York Times*, May 20, 2007.
2. Bill Gates, "A New Approach to Capitalism in the 21st Century," World Economic Forum, Davos, Switzerland, January 24, 2008. http://creativecapitalism.typepad.com/creative_capitalism/2008/06/bill-gates-crea.html.
3. Bill Joy, "Why the Future Does Not Need Us," *Wired*, April 2000.
4. Henry Thoreau, *Walden*, ed. Bill McKibben (Boston: Beacon Press, 1997), 16.
5. "Can Religion Withstand Technology?" *Closer to Truth*, PBS, http://www.pbs.org/kcet/closertotruth/transcripts/314_religiontech.pdf.
6. Jacques Ellul, *The Technological Society* (New York: Knopf, 1964), 140–41.
7. Neil Postman, *Technopoly: The Surrender of Culture to Technology* (New York: Vintage Books, 1993), 71.
8. David Tabachnick, "Techne, Technology and Tragedy," *Techne* 7, no. 3 (Spring 2004). http://scholar.lib.vt.edu/ejournals/SPT/v7n3/tabachnick.html.
9. See, for example, Anthony A. Hoekema, *Created in God's Image* (Grand Rapids: Wm. B. Eerdmans, 1986), 12; or David J. A. Clines, "The Image of God in Man," *Tyndale Bulletin* 19 (1968): 101. In 1 Corinthians 11:7, Paul writes that men were created in God's image, while women reflect the image of man. This has led some, such as Thomas Aquinas, to suggest that women hold God's image only derivatively; however, there is virtual consensus today that the Genesis statement is quite clear in linking creation in God's image with both male and female. See Kari Borresen, "The *Imago Dei*: Two Historical Contexts," *Mid-Stream* 21, no. 3 (July 1982): 359–65.
10. Gerhard Kittel, ed., *Theological Dictionary of the New Testament*, trans. Geoffrey W. Bromiley (Grand Rapids: Eerdmans, 1964), under "The Divine Likeness in the OT," by Gerhard von Rad, 391–92.

11. See I Sam 6:5, Num 33:52, II Kings 11:18, Ez 23:14. Gerhard von Rad, *Genesis: A Commentary*, trans. John H. Marks, The Old Testament Library (Philadelphia: Westminster, 1961), 56.
12. Ibid., 58.
13. Louis Pirot, ed., *Supplement, Dictionnaire de la Bible* (Paris: Letouzey, 1928), under "Ressemblance et image de Dieu," by P. E. Dion.
14. Jeffrey Tigay, *The Evolution of the Gilgamesh Epic* (Philadelphia: University of Pennsylvania Press, 1982), 153ff.
15. Von Rad, *Genesis*, 56. See also Gunnlaugur A. Jonsson, *The Image of God: Genesis 1:26–28 in a Century of Old Testament Research*, Coniectanea Biblica Old Testament Series, ed. Tryggve Mettinger and Magnus Ottosson, no. 26 (Lund: Almqvist and Wiksell, 1988), 222.
16. Gerhard von Rad, *Old Testament Theology*, vol. 1, *The Theology of Israel's Historical Traditions*, trans. D. M. G. Stalker (New York: Harper and Row, 1962), 144. See also von Rad, "Divine Likeness," 392. Lynn White was one of the first to blame the destructive tendencies of Western culture toward nature on attitudes derived from the notion that humans are radically separate from nature in that only they are created in God's image and likeness. Lynn White, "The Historical Roots of our Ecological Crisis," *Science* 155 (March 10, 1967): 1203–07. Douglas Hall makes a similar charge in *Imaging God: Dominion as Stewardship* (Grand Rapids: Eerdmans, 1986), 21–26.
17. Karl Barth, *Church Dogmatics*, vol. 3, *The Doctrine of Creation*, pt. 1, ed. G. W. Bromiley and T. F. Torrance, trans. J. W. Edwards, O. Bussey, and Harold Knight (Edinburgh: T. and T. Clark, 1958), 182.
18. Ibid., 185.
19. Karl Barth, *Table Talk*, ed. John D. Godsey (Richmond, VA: John Knox Press, 1962), 57.
20. Barth, *Church Dogmatics*, vol. 3, 186.
21. The idea of the image as an I/Thou relationship does not originate with Barth. It has its foundation in Martin Buber's work, as well as the thought of a whole host of theologians who explored diological personalism in the 1920s and 1930s, including Dietrich Bonhoeffer and Emil Brunner. Moreover, aspects of a relational image are found in the writing of the early reformers Martin Luther and John Calvin. However, Barth is the first to base his relational image on the man/woman relationship. In this, he was likely highly influenced by the thought of his secretary and companion Charlotte von Kirschbaum. For a discussion of von Kirschbaum's influence on Barth's thought, see Suzanne Selinger, *Charlotte von Kirschbaum: A Study in Biography and the History of Theology* (University Park: Pennsylvania State University, 1998).
22. Although many systematic theologians have differed sharply from Barth on the details of what constitutes authentic relationship, or whether the male–female differentiation mentioned in Genesis 1:26 is an adequate model for all human relationship, a relational understanding of humans as the image of God has become the dominant approach among systematic theologians in the mid- to late-twentieth century. Relational interpretations are found in the works of Brunner, Bonhoeffer, Gerrit Cornelis Berkouwer, Wolfhart Pannenberg, and Hans Kung, among others.
23. Barth, *Church Dogmatics*, vol. 3, 187.

24. Ibid.
25. Ibid., 216.
26. Stuart McLean, "Creation and Anthropology," in *Theology beyond Christendom: Essays on the Centenary of the Birth of Karl Barth, May 10, 1886*, ed. John Thompson, Princeton Theological Monograph Series, no. 6 (Allison Park, PA: Pickwick, 1986), 117.
27. All Qur'anic quotations are from Abdullah Yusuf Ali, *The Qur'an, Translation* (Elmhurst, NY: Tahrike Tarsile Qur'an, 2007).
28. Seyyed Hossein Nasr, "Islam and the Environmental Crisis," in *Islam and the Environment*, ed. A. R. Agwan (Delhi, India: Institute of Objective Studies, 1997), 22.
29. For a further discussion of the Amish and technology, see Donald Kraybill, *The Riddle of Amish Culture* (Baltimore: The Johns Hopkins University Press, 1989).

Chapter 2: Healing or Enhancing?

1. M. Al-Jibaly. "The Inevitable Journey, Part 1—Sickness: Regulations and Exhortations" (Arlington, TX: Al-Kitaab & As-Sunnah Publishing, 1998).
2. Heather Summerhayes Cariou, *Sixty-Five Roses* (Toronto: McArthur and Co., 2006). http://www.sixtyfiverosesthebook.com/index.php/excerpt.
3. Autism seems to be linked to abnormalities on five different chromosomes. Parkinson's disease is linked to a gene on chromosome 4; multiple sclerosis, linked to alterations in a gene on chromosome 6; and Alzheimer's disease, linked to a gene on chromosome 19.
4. This sequencing was only about 83 percent complete, with over 150,000 gaps. By 2006, it was considered essentially complete, with only 341 gaps (*Nature*, supplement 1, 2006: 7). It was accomplished both in the public and private sectors, by the publicly funded Human Genome Project led by Francis Collins, and by Celera Genomics, led by Craig Venter. Collins published his results in *Nature* on February 15, 2001; Venter published his simultaneously in *Science* (February 16, 2001).
5. The instructions needed to guide the development and activities of each of our cells are encoded in our DNA. DNA molecules are made up of two twisting strands (the double helix) of four nucleotide bases (A, T, G, and C). Each base in one strand is paired with a base on the other strand. A sequence of base pairs makes up a gene. Our genes are stored in the twenty-three pairs of chromosomes located in the nucleus of every cell. Genes direct the production of proteins, which make up the structure of our cells and carry signals from cell to cell.
6. "'Miracle' gene therapy trial halted," *New Scientist* (October 3, 2002), http://www.newscientist.com/article.ns?id=dn2878.
7. Eric Juengst, "Germ-Line Gene Therapy: Back to Basics," *Journal of Medicine and Philosophy* 16, no. 6 (December 1991): 587–92.
8. Anyone who has had a pig valve replacement for a heart valve is already an animal-human hybrid. The difference is that this hybridization is not transferrable, as would also be the case with somatic cell genetic transfer. The real issue is germline transfer, which would not only give that human being new

qualities but also make those qualities transferrable to the next generation, raising the question of whether a new species has been created.

9. Michael Sandel, "The Case against Perfection," *The Atlantic* (April 2004), http://www.theatlantic.com/doc/200404/sandel.
10. Ibid.
11. Susan Garfinkel, "Stem Cells and Parkinson's Disease," International Society for Stem Cell Research Web Site, March 31, 2005, http://www.isscr.org/public/parkinsons.htm.
12. Gretchen Vogel, "Misguided Chromosomes Foil Primate Cloning," *Science* 300 (April 11, 2003), 225–27.
13. President Bill Clinton, as reported in *Science* 276 (June 27, 1997), 1951.
14. John Paul II, *Declaration on the Production and the Scientific and Therapeutic Use of Human Embryonic Stem Cells,* August 2000, http://www.vatican.va/roman_curia/pontifical_academies/acdlife/documents/rc_pa_acdlife_doc_20000824_cellule-staminali_en.html.
15. Public funding could be used for research on the roughly sixty-five preexisting stem cell lines, but many of these lines were found to be contaminated and of little use.
16. Frank E. Young, "A Time for Restraint," *Science* 287 (February 25, 2000): 1424.
17. Ted Peters, *Playing God? Genetic Determinism and Human Freedom,* 2nd ed. (New York: Routledge, 2003), 182.
18. Stem cells from either the umbilical cord or the amniotic fluid show a greater multipotency than other adult stem cells—therefore, a bit more promise for research.
19. "Gene Therapy: Islamic Rules and Regulations," http://www.islamonline.net/servlet/Satellite?pagename=IslamOnline-English-Ask_Scholar/FatwaE/FatwaE&cid=1119503545702.
20. "Stem Cell Research in Shar'ia Perspective," http://www.islamonline.net/servlet/Satellite?pagename=IslamOnline-English-Ask_Scholar/FatwaE/FatwaE&cid=1119503545118.
21. "Cloning and Its Dangerous Impacts," http://www.islamonline.net/servlet/Satellite?pagename=IslamOnline-English-Ask_Scholar/FatwaE/FatwaE&cid=1119503544346.
22. Distribution remains a problem. Since 2003, treatment through the use of retroviral drugs has increased tenfold, yet only 1 million AIDS sufferers in sub-Saharan Africa have access to these drugs. World Health Organization and UNAIDS, "December 06 AIDS Epidemic Update," http://data.unaids.org/pub/EpiReport/2006/2006_EpiUpdate_en.pdf.
23. Peter Kramer, "There's Nothing Deep about Depression," *New York Times,* April 17, 2005.
24. Kevin Kelly, "The Ethic of Humility and the Ethics of Psychiatry," *Psychiatric Times* 15, no. 6 (June 1998), http://www.searchmedica.com/htmlresource.do?c=ps&ss=psychTimesLink&p=Convera&rid=ds3-vb:p:2001t:2569793289512:4a893fa07f0505fc:471dbd38&kw=Kevin%20Kelly.
25. Kramer, 19.
26. Eli Lilly has recently repackaged Prozac as Serafem, indicated for women suffering from PMDD, a new diagnosis that seems to be a more severe form of PMS.

Many patients who dislike the association between Prozac and depression, a mental illness, are much more comfortable taking the identical drug, reassociated with a physical illness.

27. Maurice Drury, *The Danger of Words* (Bristol, England: Thoemmes, 2003), 116–17.
28. Quoted in William James, *The Varieties of Religious Experience* (New York: McMillan, 1961), 135.
29. Gordon Allport, *The Individual and His Religion* (New York: McMillan, 1950), 73.
30. World Health Organization, http://www.who.int/mental_health/management/depression/definition/en/.
31. National Institutes of Mental Health, "Women and Depression: Discovering Hope," http://www.nimh.nih.gov/publicat/depwomenknows.cfm.
32. "Mad Melancholic Feminista: A Feminist Analysist of What It's like to Live in the Prozac Nation," http://melancholicfeminista.blogspot.com/2006/02/prozac-feminism.html.
33. Jacques Ellul, *The Technological Society* (New York: Knopf, 1964), 348.
34. Lawrence Diller, *Running on Ritalin: A Physician Reflects on Children, Society, and Performance in a Pill* (New York: Bantam, 1998), 35.
35. Edward Hallowell and John Ratey, *Driven to Distraction: Recognizing and Coping with Attention Deficit Disorder from Childhood through Adulthood* (New York: Simon and Schuster, 1994). See also the President's Council of Bioethics report, "Human Flourishing, Performance Enhancement, and Ritalin," http://www.bioethics.gov/background/humanflourish.
36. See, for example, Diller, *Running on Ritalin*, chapter 5.
37. Diller, *Running on Ritalin*, 6.
38. This is the thesis of Richard DeGrandpre, *Ritalin Nation: Rapid-fire Culture and the Transformation of Human Consciousness* (New York: Norton, 1990).
39. F. X. Castellanos and R. Tannock, "Neuroscience of Attention Deficit/Hyperactivity Disorder: The Search for Endophenotypes," *Nature Reviews: Neuroscience* 3 (2002): 617–28.
40. President's Council on Bioethics, "Human Flourishing, Performance Enhancement, and Ritalin."
41. "College Life: The Ritalin Advantage?" *New York Times News Service*, August 2005, 1.
42. Jere Longman, "The Deafening Roar of the Shrug," *New York Times*, July 29, 2007.
43. Ellul, *Technological Society*, 330.
44. A major impetus to research in prosthetics and biomechatronics is the current war in Iraq. The advent of body armor and the widespread use of landmines have led to fewer deaths among soldiers but far more amputations. It is estimated that the number of persons in the U.S. with prosthetic limbs will increase by 42 percent by 2020. Civilian need in Iraq is equally high. The Red Crescent estimates this need at over three thousand a year in the Mosul area alone. Peter Beaumont, "Amputations Bring Health Crisis to Iraq," *The Observer*, 29 July 2007, World News section, 31.
45. Mary Rose Roberts, "Johns Hopkins Unveils Proto 2," *The O & P Edge* (November 2007), http://www.oandp.com/edge/issues/articles/2007-11_02.asp.

46. National Institute on Deafness and Other Communication Disorders, "Cochlear Implants," http://www.nidcd.nih.gov/health/hearing/coch.asp.
47. Steffen Rosahl, "Vanishing Senses—Restoration of Sensory Function by Electronic Implants," *Poiesis Prax* (2004) 2: 289.
48. Kwabena Boahen, "Neuromorphic Microchips," *Scientific American* (May 2005), 58–59. Very little progress has been made in the enhancement of other senses, such as touch, smell, or taste.
49. Helen Mayberg, "Deep Brain Stimulation for Treatment-Resistant Depression," *Neuron* 45 (March 2005): 651–60.
50. For a review of the state of motor system biomechatronics, see P. H. Veltink, H. F. Koopman, F. C. van der Helm, and A. V. Nene, "Biomechatronics—Assisting the Impaired Motor System," *Archives of Physiology and Biochemistry* 109, no. 1 (2001): 1–9.
51. See the websites of the Biomechatronics Group at MIT and the Berkeley Robotics Laboratory for more details: http://biomech.media.mit.edu/research/pro3_3.htm and http://bleex.me.berkeley.edu/bleex.htm.
52. Sarah Yang, "UC Berkeley researchers developing robotic exoskeleton that can enhance human strength and endurance," UC Berkeley press release, March 3, 2004, http://www.berkeley.edu/news/media/releases/2004/03/03_exo.shtml.
53. Rosahl, 287.
54. Ibid., 293.
55. Kevin Warwick, professor of Cybernetics at the University of Reading, has implanted a chip in his arm that allows him remotely to control the movement of a robotic arm through nerve signals. However, he is one of few interested in such an implant. I am reminded of *Star Trek: The Next Generation*, episode 35 (1987), "The Measure of a Man," in which Lt. Commander Data notes that Command Officer Geordie La Forge's bionic eyes perform functions far beyond the human eye, yet few humans seem to be wishing to replace their biological eyes with bionic ones.
56. Quoted in Jere Longman, "An Amputee Sprinter: Is He Disabled or Too-Abled?" *The New York Times*, May 15, 2007.
57. The U.S. Food and Drug Administration estimates that 1.5 million Americans use a ventilator each year. Paula Kurtzweil, "When Machines Do the Breathing," *FDA Consumer* (September–October 1999), http://www.fda.gov/FDAC/features/1999/599_vent.html.
58. Kevin O'Rourke, "Ethical Criteria for Removing Life Support," March 1993, http://www.domcentral.org/study/kor/93031407.htm.
59. Many patients suffering a critical illness also experience acute depression, and this possibility must be taken into account when a patient requests withdrawal of life support. Noreen Henig, John Faul, and Thomas Raffin, "Biomedical Ethics and the Withdrawal of Advanced Life Support," *Annual Review of Medicine* 52 (2001): 79–92.
60. Ibid.
61. According to a report by the Multi-Society Task force on PVS concerning adults who suffered PVS as a result of a traumatic injury, one year later, 33 percent are dead, 14 percent remain in PVS, and 52 percent exhibit some degree of recovery. For those who remain in PVS, the average life expectancy is two to

five years, though some have lived as long as forty years in this condition. "Medical Aspects of the Persistent Vegetative State," *New England Journal of Medicine* 330 (May 1994): 1499–1508; (June 1994): 1572–79.

62. It was estimated over a decade ago that the care of a PVS patient for one year can cost anywhere from $97,000 to $180,000. Donal O'Mathuna, "Responding to Patients in a Persistive Vegetative State," *Philosophia Christi* 19, no. 2 (Fall 1996), http://www.xenos.org/ministries/crossroads/donal/pvs.htm.
63. Robert Frank, "Weighing the True Costs and Benefits in a Matter of Life and Death," *The New York Times*, January 19, 2006.
64. Daniel Sulmasy, "Preserving Life? The Vatican & PVS," *Commonweal* (7 December 2007): 16.
65. Ibid.
66. IMANA Ethics Committee, "Islamic Medical Ethics: The IMANA Perspective," http://data.memberclicks.com/site/imana/IMANAEthicsPaperPart 1.pdf.
67. Stanley Hauerwas, *Truthfulness and Tragedy: Further Investigations in Christian Ethics* (Notre Dame, IN: University of Notre Dame Press, 1977), 130.

Chapter 3: Cyberspace on Our Minds

1. Pope Paul VI, "Evangelii Nuntiandi" (Dec. 8, 1975), http://www.vatican.va/holy_father/paul_vi/apost_exhortations/documents/hf_p-vi_exh_19751208_evangelii-nuntiandi_en.html.
2. Abdullah Suhrawardy, *The Sayings of Mohammad* (London: Constable, 1910), http://muslim-canada.org/hadiths.html.
3. John McCarthy, "Some Expert Systems Need Common Sense," *Annals of the New York Academy of Sciences* 426 (1984): 129–35.
4. Friedrich Nietzsche, *Thus Spake Zarathustra*, trans. Walter Kauffman (New York: Viking, 1954), 146.
5. John Haugeland, "What is Mind Design," in *Mind Design II: Philosophy, Psychology, Artificial Intelligence*, ed. John Haugeland (Cambridge, MA: MIT, 1997), 26.
6. Benjamin Libet, "Do We Have Free Will?" *Journal of Consciousness Studies* 6, nos. 8–9 (1999): 47.
7. The main exception to this is the computer Dave in the film *2001: A Space Odyssey*.
8. For descriptions of a variety of mobile robots developed at MIT, see Rodney Brooks, "Elephants Don't Play Chess," *Robotics and Autonomous Systems* 6 (1990): 3–15.
9. Alan Turing, "Computing Machinery and Intelligence," in Haugeland, 29–32.
10. Ibid., 38.
11. While most in the AI community accept the Turing Test as sufficient, an opposing view can be found in John Searle, "Minds, Brains, and Programs," *The Behavioral and Brain Sciences* 3 (1980): 417–24.
12. Terry Winograd and Fernando Flores, *Understanding Computers and Cognition: A New Foundation for Design* (Norwood, NJ: Ablex, 1986; reprint, Reading, MA: Addison-Wesley, 1991), 68.
13. J. Z. Young, *Programs of the Brain* (Oxford: Oxford University Press, 1978), 194.

14. Reported in Jonah Lehrer, "Hearts and Mind," *The Boston Globe*, April 29, 2007.
15. Marvin Minsky, *The Society of Mind* (New York: Simon and Schuster, 1985).
16. Rosalind Picard, *Affective Computing* (Cambridge, MA: MIT Press, 1997), chapter 2.
17. Neal Stephenson, *Snow Crash* (New York: Bantam, 2000), 33.
18. Ray Kurzweil, *The Age of Spiritual Machines: When Computers Exceed Human Intelligence* (New York: Penguin, 1999), chapter 6.
19. Frank Tipler, *The Physics of Immortality: Modern Cosmology, God, and the Resurrection of the Dead* (New York: Doubleday, 1995).
20. Francis Crick, *The Astonishing Hypothesis: The Scientific Search for the Soul* (New York: Charles Scribner's Sons, 1994), 3.
21. Donald MacKay, *Behind the Eye* (Oxford: Basil Blackwell, 1991), 260.
22. Reinhold Niebuhr, *The Nature and Destiny of Man: A Christian Interpretation*, vol. 1, *Human Nature*, with intro. by Robin Lovin, Library of Theological Ethics (Louisville: Westminster John Knox, 1996), 178–79.
23. Tipler, 255. While this is not the place for an extended feminist critique, one cannot help but notice that the proponents of cybernetic immortality and artificial intelligence are overwhelmingly male. Women remain in their speculations as objects of desire yet are stripped of their reproductive role. Disembodied sexual experience, in the form of pornography, is, of course, a staple of the Internet.
24. Jaron Lanier, "One-Half of a Manifesto," *Wired* 8, no. 12 (December 2000): 4.
25. Joanna Bryson and Phil Kime, "Just Another Artifact," http://www.cs.bath.ac.uk/~jjb/web/aiethics98.html.
26. David Walsh, Douglas Gentile, Jeremy Gieske, Monica Walsh, and Emily Chasko, *Mediawise Video Game Report Card*, National Institute on Media and the Family (December 8, 2003): 3.
27. Craig Anderson, Douglas Gentile, and Katherine Buckley, "Study 3: Longitudinal study with elementary school students," in *Violent Video Game Effects on Children and Adolescents* (New York: Oxford University Press, 2007), 95–119.
28. Douglas Gentile, Paul Lynch, Jennifer Linder, and David Walsh, "The Effects of Violent Video Game Habits on Adolescent Hostility, Aggressive Behaviors, and School Performance," *Journal of Adolescence* 27 (2004): 5–22.
29. David Gale, "Research Links Violence to Video Games," *BYU NewsNet*, 29 April 2003.
30. Akira Sakamoto, "Video Game Use and the Development of Sociocognitive Abilities in Children: Three Surveys of Elementary School Children," *Journal of Applied Social Psychology* 24 (1994): 21–42.
31. Anderson et al., *Violent Video Game Effects*, 61–119.
32. Craig Anderson, "Violent Video Games: Myths, Facts, and Unanswered Questions," American Psychological Association On-line, http://www.apa.org/science/psa/sb-andersonprt.html.
33. David Grossman, lecture, St. John's University, Collegeville, MN, February 9, 2005.
34. Wesley Clark, "Iraq: What Went Wrong," *The New York Review of Books*, October 23, 2003, 52–54.

35. Kate Zernike, "Violent Crime in Cities Shows Sharp Surge," *The New York Times*, March 9, 2007.
36. C. Shawn Green and Daphne Bavelier, "Action Video Game Modifies Visual Selective Attention," *Nature* 423 (May 29, 2003): 534–36.
37. Steven Johnson, "This is Your Brain on Video Games," *Discover*, July 9, 2007, http://discovermagazine.com/2007/brain/video-games.
38. Mary Ellen Shay, "Teen-Rated Video Games Loaded with Violence," press release, Children's Hospital Boston, March 11, 2004.
39. Gayle Hanson, "The Violent World of Video Games," *Insight on the News* (June 28, 1999): 15.
40. Quoted in "No Hiding Place," *Economist* 373, no. 8399 (October 30, 2004): 83.
41. Michael Brody, "Playing with Death," *Brown University Child & Adolescent Behavior Letter* (November 2000): 8.
42. Espen Aarseth, *Cybertext: Perspectives on Ergodic Literature* (Baltimore: Johns Hopkins, 1997), 162.
43. The American Medical Association discussed the possibility of designating video-game addiction as a mental disorder at their 2007 meeting but tabled the proposal, citing the need for more research. It will be reconsidered in 2012.
44. Michael Benedikt, "Introduction," *Cyberspace: First Steps*, ed. Michael Benedikt (Cambridge, MA: MIT, 1991), 14–15.
45. It is sometimes also called "sniper mode."
46. Tim Boucher, "God-mode: Video Games, Virtual Reality & Religion," http://www.timboucher.com/investigations/god_mode_maya_games.php (essay no longer available).
47. Carl Pope, "The Forgotten Family Value," *Sierra* (November/December 2000): 16.
48. Heather Wax, "In a Virtual World, What Happens When the Bad Guy Wins?" *Research News and Opportunities in Science and Theology* (October 2003): 35.
49. Simulations are among the few video games played extensively by girls. Video games aimed strictly at a female demographic have been singularly unsuccessful. One can see why; titles include *Barbie's Fashion Designer*, *Let's Talk about Me*, and *Rockett's Secret Invitation*, in which a girl navigates through the social cliques of a virtual junior high. Beyond simulations, the video world is a somewhat hostile place for women. Rape scenes, prostitution, full nudity, and disembodied breasts are common fare.
50. David Grossman, *On Combat*, http://www.killology.com/on_combat_ch7.htm.
51. Eugene Provenzo, testimony before the United States Senate Joint Hearing of the Judiciary Subcommittee on Juvenile Justice and the Government Affairs Subcommittee on Regulation and Government Information on the issue of violence in video games, December 9, 1993.
52. Christopher Maag, "When the Bullies Turned Faceless," *The New York Times*, December 16, 2007.
53. Pete Cashmore, "MySpace Hits 100 Million Accounts," *Mashable: Social Networking News*, http://mashable.com/2006/08/09/myspace-hits-100-million-accounts/; Facebook website, http://www.facebook.com/press/info.php?statistics.

54. Amanda Lenhart, Mary Madden, Alexandra MacGill, and Aaron Smith, "Teens and Social Media," Pew Internet and American Life Project, http://www.pewinternet.org/PPF/r/230/report_display.asp.
55. Quentin Schulze, *Habits of the High-Tech Heart: Living Virtuously in the Information Age* (Grand Rapids, MI: Baker Academic, 2002), 174.
56. Marshall McLuhan, *Understanding Media: The Extensions of Man* (Cambridge, MA: MIT Press, 1994), 7.
57. Christine Rosen, "Virtual Friendship and the New Narcissism," *The New Atlantis* (Summer 2007): 24–25.
58. Aaron Smith, "Teens and Online Stranger Contact," Pew Internet and American Life Project, http://www.pewinternet.org/PPF/r/223/report_display.asp.
59. Ibid.
60. David Eberhardt, "Should Institutions Respect Students' On-line Privacy on Facebook?" *Journal of College and Character* (October 4, 2005), http://www.collegevalues.org/seereview.cfm?id=1529.
61. Rushworth Kidder, "The Hypocrisy of MySpace," *Ethics Newsline* (July 10, 2006), http://www.globalethics.org/newsline/2006/07/10/the-hypocrisy-of-myspace/.
62. Francis Fukuyama, "The Great Disruption: Human Nature and the Reconstitution of Social Order," *The Atlantic Monthly* (May 1999), 55–80.
63. Christopher Sanders, Tiffany Field, Miguel Diego, and Michelle Kaplan, "The Relationship of Internet Use to Depression and Social Isolation among Adolescents," *Adolescence* (Summer 2000): 237–42.
64. John Cassidy, "Me Media," *The New Yorker*, 15 May 2006, http://www.newyorker.com/archive/2006/05/15/060515fa_fact_cassidy?currentPage=all.
65. Robert Bellah, *Habits of the Heart: Individualism and Commitment in American Life* (Berkeley: University of California Press, 1985), 71.
66. Schulze, *Habits of the High-Tech Heart*, 182.
67. Cassidy, "Me Media."
68. Eric Gwinn, "Narcissism on the Internet Isn't Risk-free," *Chicago Tribune*, March 16, 2007.
69. Rosen, "Virtual Friendship and the New Narcissism," 27.
70. Karl Barth, *Church Dogmatics*, ed. G. W. Bromiley and T. F. Torrance, trans. J. W. Edwards, O. Bussey, and H. Knight (Edinburgh: T. and T. Clark, 1958), 250–63.
71. Rosen, "Virtual Friendship and the New Narcissism," 31.
72. Lorne Dawson, "Doing Religion in Cyberspace: The Promise and the Perils," *Bulletin of the Council of Societies for the Study of Religion* 30, no. 1 (February 2001): 3–8.
73. Michael Benedikt, "Cyberspace: Some Proposals," in *Cyberspace: First Steps* (Cambridge: MIT, 1991), 122.
74. Jennifer Cobb, *Cybergrace: The Search for God in the Digital World* (New York: Crown, 1998), 199.
75. David Porush, "Hacking the Brainstem: Postmodern Metaphysics and Stephenson's *Snow Crash*," in *Virtual Realities and Their Discontents*, ed. Robert Markley, 107–41 (Baltimore: Johns Hopkins, 1996), 126.
76. John Chryssavgis, "Self-Image and World Image: Ecological Insights from

Icons, Liturgy, and Asceticism," *Christian Spirituality Bulletin* (Spring/Summer 2000): 15.

77. Richard Thomas, personal e-mail, February 5, 2001.
78. Nicole Stenger, "Mind Is a Leaking Rainbow," in *Cyberspace: First Steps*, 53–54.
79. Quoted in Jeff Zaleski, *The Soul of Cyberspace: How the New Technology Is Changing Our Lives* (New York: HarperCollins, 1997), 142.
80. Michael Heim, "The Erotic Ontology of Cyberspace" in *Cyberspace: First Steps*, 76.
81. Canadian artist Char Davies is one of the few programmers working in virtual reality who has deliberately chosen to embody beauty as one of the principles in her work. Her program "Osmose" immerses the user in a luminescent world both like and unlike the natural world and has been described as inducing a state of meditative peacefulness.
82. Gennadios Limouris, "The Microcosm and Macrocosm of the Icon: Theology, Spirituality and Worship in Colour," in *Icons: Windows on Eternity*, ed. Gennadios Limouris, Faith and Order Paper 147 (Geneva: World Council of Churches, 1990), 117.
83. Chrissavgis, "Self-Image and World Image," 19.
84. Janet Murray, *Hamlet on the Holodeck: The Future of Narrative in Cyberspace* (New York: Free Press, 1997), 126.

Chapter 4: The New Alchemy

1. Martin Luther, *Table Talk*, trans. William Hazlitt (Philadelphia: The Lutheran Publication Society, 2004), 463.
2. Lynn White, "The Historical Roots of Our Ecological Crisis," *Science* 155, 10 March 1967, 1203–7.
3. Robin Henig, "Our Silver-Coated Future," *OnEarth*, Fall 2007, 24.
4. Richard Feynman, "There's Plenty of Room at the Bottom," annual meeting of the American Physical Society, December 29, 1959, http://www.zyvex.com/nanotech/feynman.html.
5. Gabriel Silva, "Introduction to Nanotechnology and Its Applications to Medicine," *Surgical Neurology* 64 (2004): 216.
6. Patel-Predd, Prachi, "Buckyballs with a Surprise," *MIT Technology Review* (1 November 2006), http://www.technologyreview.com/read_article.aspx?id=17 681&ch=nanotech.
7. Buckyballs surpass diamonds and graphite in strength and hardness. Their inventors, Robert Curl, Richard Smalley, and Harold Kroto, received the Nobel Prize in chemistry in 1996.
8. Albert Fert and Peter Grundberg received the 2007 Nobel Prize in physics for this advance.
9. Henig, "Our Silver-Coated Future," 24.
10. James Thrall, "Nanotechnology and Medicine," *Radiology* 230 (2004): 317.
11. "Water-related Diseases," World Health Organization, http://www.who.int/water_sanitation_health/diseases/diarrhoea/en/.
12. Andrew Maynard outlined five difficulties for risk assessment to the President's Council on Bioethics. These included the difficulty of measuring exposure to

nanomaterials, of testing these materials for toxicity, and of needing to look at the entire life cycle of nanoparticles, including their half-life in the environment. Transcript, "Nanotechnology: Benefits and Risks," President's Council on Bioethics (July 23, 2007), 8.

13. Ibid., 19.
14. http://www.eande.tv/transcripts/?date=120505#transcript.
15. Bill Joy, "Why the Future Does Not Need Us," *Wired* 8.04 (April 2000), 61.
16. The development of "gray goo" would be extremely unlikely. Such a goo would need to be mobile, able to live off the biosphere (air or water), and self-replicating. Nature itself, with all the tools of evolution, has been unable to produce such a product. It is unlikely that human ingenuity would do so in the foreseeable future.
17. Eric Drexler, *Engines of Creation: The Coming Era of Nanotechnology* (New York: Anchor, 1986), reprinted on the web at http://www.e-drexler.com/d/06/00/EOC/EOC_Table_of_Contents.html.
18. Adam Keiper, "The Nanotechnology Revolution," *The New Atlantis*, Summer 2003, 5. Online at http://www.thenewatlantis.com/archive/2/keiperprint.htm.
19. President's Council on Bioethics, "Nanotechnology: Benefits and Risks," 12.
20. Ray Kurzweil, "The Next Frontier," *Science and Spirit*, November/December 2005, 69.
21. Reinhold Niebuhr, *The Nature and Destiny of Man*, vol. 1 (Louisville, KY: Westminster John Knox, 1996), 163.
22. Ibid., 177.
23. Ibid., 124–25.
24. Ibid., 298.
25. Robert A. Frietas Jr., "Nanomedicine," http://www.foresight.org/Nanomedicine/.
26. Monsanto 2006 Report, http://www.monsanto.com/monsanto/layout/our_pledge/2006_report_pdf.asp (accessed March 2008; report no longer available).
27. UN Food and Agriculture Association, "The number of hungry people is increasing by 4 million per year," http://www.worldhunger.org/articles/06/global/sofi_2006.htm.
28. "What is Hunger?" United Nations World Food Programme, http://www.wfp.org/aboutwfp/introduction/hunger_what.asp?section=1&sub_section=1.
29. Andy Coghlan, "New 'Golden Rice' Carries Far More Vitamin," *New Scientist*, March 27, 2005.
30. The U.S. and Argentina account for almost 90 percent of GM crop production, with much of the remaining production in Canada and China.
31. USDA Economic Research Service Report, July 2, 2008, "Adoption of Genetically Engineered Crops in the U.S.," http://www.ers.usda.gov/Data/BiotechCrops/.
32. A look at the website for the Council for Biotechnology Information, a public-relations effort by seven biotechnology corporations, including Monsanto, shows a decided emphasis on the need for GM crops in the third world. http://www.whybiotech.com/index.asp?id=4573
33. ISAAA Brief 35-2006, http://www.isaaa.org/resources/publications/briefs/35/default.html/. See also Suwen Pan et al. "The Impact of India's Cotton Yield

on U.S. and World Cotton Markets," Cotton Economics Research Institute Briefing, http://www.aaec.ttu.edu/ceri/NewPolicy/Publications/BriefingPapers/ImpactsIndiaYieldonUS.pdf.

34. "Six More Farmers Commit Suicide," *The Hindu*, 5 July 2007. See also Fred De Sam Lazaro, *The Dying Fields*, PBS Wideangle Documentary, August 2007, http://www.pbs.org/wnet/wideangle/episodes/the-dying-fields/introduction/967/.
35. ISAAA Brief 35-2006.
36. And the crop need not entirely fail. One Vidarbha farmer reports spending $1,250 on seeds, $600 on fertilizer, and more on pesticides in 2006, yet his crop brought in only $1,200. De Sam Lazaro, *The Dying Fields*.
37. The Indian government does have programs to subsidize seed and fertilizer but makes no direct cash payments to cotton farmers.
38. Peter Rosset, "Could Genetic Engineering End Hunger?" in *Engineering the Farm: Ethical and Social Aspects of Agricultural Biotechnology*, ed. Britt Bailey and Marc Lappe (Washington, D.C.: Island Press, 2002), 90–91.
39. Laura Spinney, "Biotechnology in Crops: Issues for the Developing World," *Report for Oxfam International*, May 1998, http://www.actionbioscience.org/biotech/oxfam_spinney.html.
40. Celia Deane-Drummond, "Genetic Engineering for the Environment: Ethical Implications of the Biotechnology Revolution," *The Heythrop Journal* 36 (July 1995): 309.
41. Richard Lewontin, *It Ain't Necessarily So: The Dream of the Human Genome and Other Illusions* (New York: New York Review of Books, 2001), 366.
42. Debbie Barker, "The Rise and Predictable Fall of Globalized Industrial Agriculture," *International Forum on Globalization Report*, 2007, http://www.ifg.org/pdf/ag%20report.pdf.
43. "Basmati Rice Update," Rural Advancement Foundation International, January 2000, http://www.biotech-info.net/basmati_rice.html.
44. Quoted in Joff Wild, "The Future for Patents on Life," Thomson Scientific, January 2002.
45. A famous case is that of Perry Schmeising in Saskatchewan, who was found to have a field of over 95 percent Roundup Ready soybeans without purchase of the seeds from Monsanto.
46. Lori Andrews, "Patents, Plants, and People," in *Engineering the Farm*, 72.
47. As of 2002, patent applications had been filed on over 125,000 genes or genetic sequences. Ibid., 73.
48. Edmund Andrews, "Company News: Religious Leaders Prepare to Fight Patents on Genes," *The New York Times*, May 13, 1995.
49. Ronald Cole-Turner, "The Theological Status of DNA: A Contribution to the Debate over Gene Patenting," in *Perspectives on Genetic Patenting: Religion, Science and Industry in Dialogue*, ed. Audrey Chapman (Washington, D.C.: American Association for the Advancement of Science, 1999), 152.
50. Quoted in Mark Hanson, "Religious Voices in Biotechnology: The Case of Gene Patenting," in *Perspectives on Genetic Patenting*, 90.
51. Quoted in Richard Black, "Growing pains of India's GM revolution," *BBC News*, August 16, 2007.

52. Quoted in Trey Popp, "God and the New Foodstuffs," *Science and Spirit*, March 2006).
53. "Climate change could intensify hunger risk in developing world, UN official says," UN News Centre, September 6, 2007, http://www.un.org/apps/news/story.asp?NewsID=23446&Cr=climate&Cr1=change.
54. David Lobell and Christopher Field, "Global Scale Climate–Crop Yield Relationships and the Impacts of Recent Warming," *Environmental Research Letters* 2 (January–March 2007).
55. Daniel Yergin, *The Prize: The Epic Quest for Oil, Money, and Power* (New York: Simon and Schuster, 1992), 13–14.
56. John H. Wood, Gary R. Long, and David F. Morehouse, "Long-Term World Oil Supply Scenarios," http://www.eia.doe.gov/pub/oil_gas/petroleum/feature_articles/2004/worldoilsupply/oilsupply04.html.
57. Princeton geologist Kenneth Deffeyes wrote in 2001, "There is nothing plausible that could postpone the peak until 2009." Quoted in Joseph Romm, *Hell and High Water: Global Warming—the Solution and the Politics—and What We Should Do* (New York: Morrow, 2007), 180.
58. Neil King and Peter Fritsch, "Energy Watchdog Warns of Oil Production Crunch," *The Wall Street Journal*, May 22, 2008.
59. Quoted in Richard Heinberg, *Powerdown: Options and Actions for a Post-Carbon World* (Gabriola Island, Canada: New Society, 2004), 17.
60. This also leads to increased pollution. Scientists have found that the smoke from the 100 million cooking stoves in India's villages is leading to near constant smog from the Himalayas to the Maldives during the winter months. There is concern that this smog could disrupt India's monsoon cycle. Fred Pearce, *With Speed and Violence: Why Scientists Fear Tipping Points in Climate Change* (Boston: Beacon, 2007), 115–19.
61. Norway seems to be the exception to the rule. Its economy is highly oil based yet has not shown any of the deleterious social and political effects we see in developing countries. Russia, however, is another story, one that goes beyond the scope of this section.
62. Jim Vallette and Steve Kretzmann, "The Energy Tug of War: The Winners and Losers of World Bank Fossil Fuel Finance," Institute for Policy Studies, April 2004.
63. See, for example, "Chad's Oil Riches, Meant for Poor, are Diverted," *The New York Times*, February 18, 2006.
64. "President Commemorates 60th Anniversary of V-J Day," http://www.whitehouse.gov/news/releases/2005/08/20050830-1.html.
65. This raises an interesting question about the war on terrorism. Osama bin Laden's stated objective is the withdrawal of American influence in Saudi Arabia and, more recently, the entire Middle East. Given the strategic importance of the region's oil, America will not be able to withdraw and will seek increasing influence in the region as demand for oil increases over supply. Thus, the war on terrorism is intrinsically linked to our dependence on oil.
66. Mark Clayton, "Global Boom in Coal Power—and Emissions," *The Christian Science Monitor*, March 22, 2007.
67. Ibid.

68. International Energy Agency, *World Energy Outlook 2006*, Executive Summary, http://www.iea.org/textbase/nppdf/free/2006/weo2006.pdf.
69. "Report of Working Group 1, Summary for Policymakers," Intergovernmental Panel on Climate Change, 3–5.
70. Ibid., 16–18.
71. Ice in the Arctic is already at new lows this summer; Russia and Canada are both anticipating an ice-free Arctic and jockeying for position. See Andrew Revkin, "Analysts See 'Simply Incredible' Shrinking of Floating Ice in the Arctic," *The New York Times*, August 10, 2007.
72. "Bush Sees Green Reasons for Nuclear Power," *MSNBC*, June 22, 2005, http://www.msnbc.msn.com/id/8315963/.
73. Quoted in Romm, *Hell and High Water*, 175.
74. http://www.ocrwm.doe.gov/ym_repository/about_project/waste_explained howmuch.shtml.
75. Ian Barbour, *Ethics in an Age of Technology: The Gifford Lectures, 1989–1991*, vol. 2 (New York: HarperCollins, 1993), 127–28.
76. Hermann Scheer, *Energy Autonomy: The Economic, Social and Technological Case for Renewable Energy* (London: Earthscan, 2007), 100.
77. Alexander Farrell et al., "Ethanol Can Contribute to Energy and Environmental Goals," *Science* (January 27, 2006).
78. "Experts Differ About Ethanol-Water Usage," *The New York Times*, August 10, 2007.
79. James McKinley, "Cost of Corn Soars, Forcing Mexico to Set Price Limits," *The New York Times*, January 19, 2007.
80. Quoted in John Schoen, "Ethanol: Boom or Boondoggle?" MSNBC, 19 November 2006, http://www.msnbc.msn.com/id/3540967.
81. U.S. Department of Energy, http://www.eere.energy.gov/.
82. Scheer, *Energy Autonomy*, 131.
83. Peter Kareiva et al., "Domesticated Nature; Shaping Landscapes and Ecosystems for Human Welfare," *Science* 316 (June 29, 2007): 1866.
84. Elizabeth Kolbert, *Field Notes from a Catastrophe: Man, Nature, and Climate Change* (New York: Bloomsbury, 2006), 187.

Chapter 5: Technology Goes Global

1. Manuel Castells, "Information Technology, Globalization and Social Development," UNRISD Discussion Paper no. 114 (September 1999): 4. http://www.unrisd.org/unrisd/website/document.nsf/ab82a6805797760f80256b4f005da1ab/f270e0c066f3de7780256b67005b728c/$FILE/dp114.pdf.
2. Carlo Fonseka, "Globalization and Universality of Humankind," *Dialogue* 24 (1997): 2.
3. Nicholas Kristof, "Extended Forecast: Bloodshed," *The New York Times*, April 13, 2008.
4. Giovanna Prennushi and Stephen Browne, "Cutting Poverty," *OECD Observer*, 2008, http://www.oecdobserver.org/news/fullstory.php/aid/369/Cutting_poverty.html.
5. Castells, "Information Technology, Globalization and Social Development," 8.

6. Ibid., 8–9.
7. World Stem Cell Map, Minnesota Biomedical and Bioscience Network, University of Minnesota, http://www.mbbnet.umn.edu/scmap.html.
8. Kirk Biglione, "The Secret to Japan's Robot Dominance," *Planet Tokyo*, January 24, 2006, http://www.planettokyo.com/news/index.cfm/fuseaction/story/ID/36/.
9. C. L. Richard, "Why Biotech Foods Are Kosher," *Revista Da Cumnidade Judaica Brasileira* (April 2000), http://www.agbioworld.org/biotech-info/religion/kosher.html.
10. Michael Green, "Why GM Food Isn't Kosher," *Jewish Chronicle* (April 27, 2007), 43.
11. Anthony Giddens, *Runaway World: How Globalization Is Reshaping Our Lives*, 2d ed. (London: Profile Books, 2002), 61–66.
12. Diana Eck, *New Religious America*, quoted in S. Wesley Ariarajah, "Religious Diversity and Interfaith Relations in a Global Age," 2007 International Conference on Religion and Culture, Institute of Religion, Culture and Peace, Chiang Mai, Thailand, http://isrc.payap.ac.th/document/speeches/speech01.pdf.
13. Pew Forum on Religion and Public Life, "U.S. Religious Landscape Survey," http://religions.pewforum.org/reports.
14. William Cantwell Smith, *Towards a World Theology* (London: Macmillan, 1981).
15. Ariarajah, "Religious Diversity and Interfaith Relations in a Global Age."
16. Giddens, *Runaway World*, 64.
17. Alfred Schutze, *The Enigma of Evil*, quoted in Andrew Kimbrell, "Cold Evil: Technology and Modern Ethics," Twentieth Annual E. F. Schumacher Lecture, October 2000, http://www.schumachersociety.org/publications/kimbrell_00.html.
18. Emmanuel Levinas, *Ethics and Infinity: Conversations with Philippe Nemo* (Pittsburgh: Duquesne University Press, 1985), 87.
19. Kimbrell, "Cold Evil: Technology and Modern Ethics."
20. "Vatican Bishop Points to Modern Social Sins," Catholic News Agency, 11 March 2008, http://www.catholicnewsagency.com/new.php?n=12031.
21. Mohandas Gandhi, "The Seven Social Sins," *Young India* (October 22, 1925).
22. Swami Jitatmananda, "Bioethics for Science and Technology: A Hindu Perspective," *Prabuddah Bharata* (February 2005). http://www.eng.vedanta.ru/library/prabuddha_bharata/Feb2005_bioethics_for_science_and_technology.php.
23. "The Islamic Perspective on the Environmental Crisis: Seyyed Hossein Nasr in Conversation with Muzaffar Iqbal," http://www.thefreelibrary.com/The+Islamic+perspective+on+the+environmental+crisis:+Seyyed+Hossein...-a0164596587.
24. Ibid.
25. Reinhold Niebuhr, *The Nature and Destiny of Man*, vol. 1 (Louisville, KY: Westminster John Knox, 1996), 321.

Appendix

1. See http://www.thewords.com/articles/ellul76quest.htm.

Index